Props, Fans, & Pumps

Design & Performance

by **D. James Benton**

Preface

Propellers, fans, and pumps are an integral part of many machines of industry, transportation, and modern living. The designs are as diverse as the applications, but the theory behind how these devices work is the same. In this text we will consider nuances of design, such a prop cupping, trailing edge truncation, and blade curling. We will derive and apply the formulas in order to design and evaluate the performance of these fascinating devices. All of the software described herein is available free online.

All of the examples contained in this book,
(as well as a lot of free programs) are available at...
https://www.dudleybenton.altervista.org/software/index.html

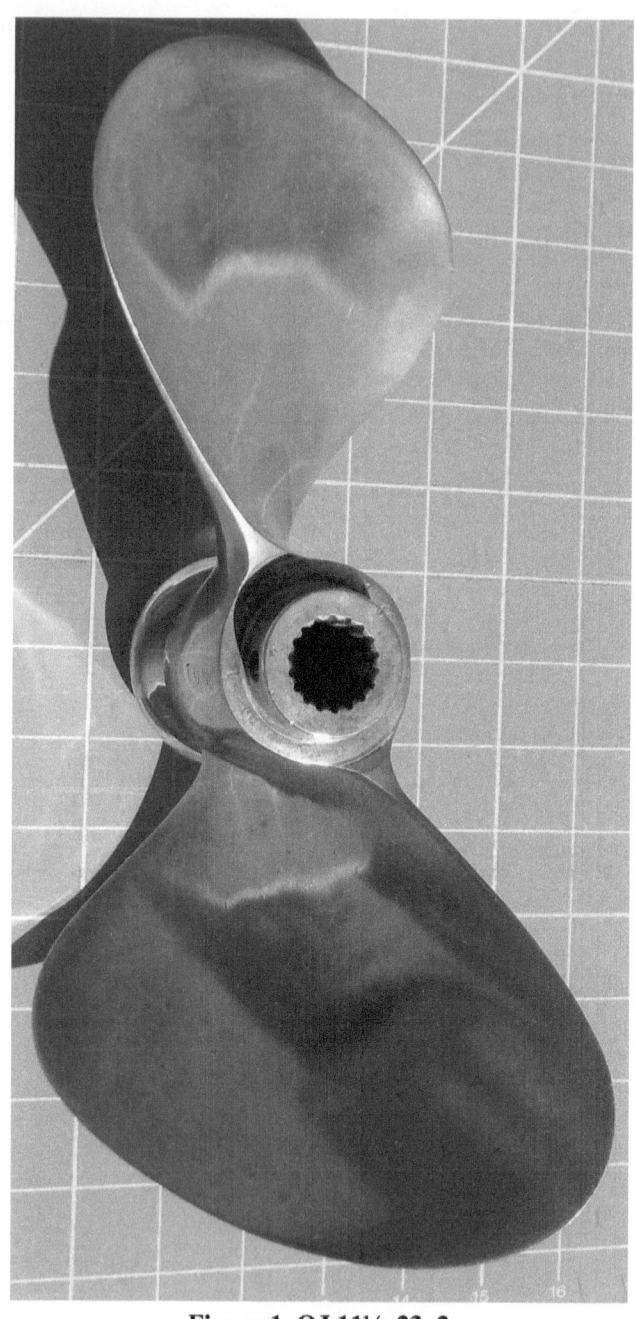

Figure 1. OJ 11½x23x2

Table of Contents

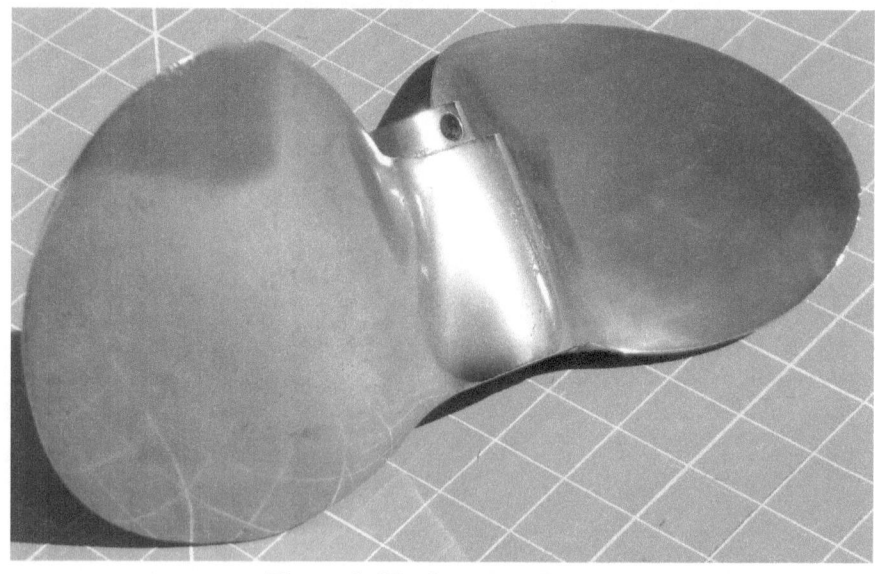

Figure 2. Three for a Super-BP

Figure 3. Elephant Ear Design

Chapter 1. Basic Theory

We begin with the most basic theory for fans. Consider the following diagram showing the ideal slipstream around a propeller:

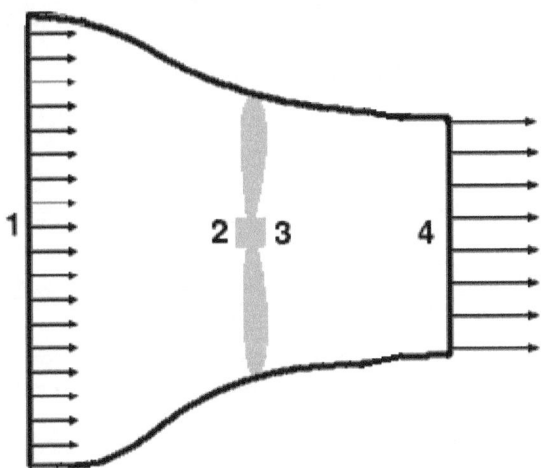

Figure 4. Propeller and Slipstream Boundary

The boundary in this figure may or may not correspond to a physical surface, such as a shroud. In practice, for a propeller spinning in an open fluid, the boundary be difficult or even impossible to delineate. The conservation of mass can be expressed as follows, at least conceptually:

$$\rho_1 V_1 A_1 = \rho_2 V_2 A_2 = \rho_3 V_3 A_3 = \rho_4 V_4 A_4 \tag{1.1}$$

If the fluid were water, the density at each of the four points would be essentially the same. In air, as long as the velocities are not too large and the pressure change across the propeller too great, this approximation would also be reasonable, which reduces the equation to:

$$V_1 A_1 = V_2 A_2 = V_3 A_3 = V_4 A_4 \tag{1.2}$$

If we reckon that the distance between points 2 and 3 across the propeller is small compared to the distance to either 1 or 4, then the area on either side of the propeller is the same.

$$A_2 = A_3 \tag{1.3}$$

1

When the slipstream starts (point 1) and when it ends (point 4) is somewhat arbitrary. This is like answering the question, "How close (or far) is close (or far) enough?" Rather than impose one arbitrary measure, such as might be in line with boundary layer theory (when the velocity is 99% of the free stream), we will simply define a convenient relationship between the four velocities. This says nothing about the distance from 1 to 2 or from 3 to 4, which we don't require a this point.

$$V_2 = V_3 = \frac{V_1 + V_4}{2} \tag{1.4}$$

We can also deduce a relationship between the areas:

$$A_2 = A_3 = 2\frac{A_1 A_4}{A_1 + A_4} \tag{1.5}$$

Assuming a round cross section proportional to diameter squared:

$$D_2^2 = D_3^2 = 2\frac{D_1^2 D_4^2}{D_1^2 + D_4^2} \tag{1.6}$$

As we balance forces along the centerline (not the only forces, but often the only significant net forces), we conclude that the thrust is equal to the pressure across the propeller times the area:

$$F = A_2\left(P_3 - P_2\right) \tag{1.7}$$

We know from observations, including the fact that the pressure far away from the propeller is independent of position, that the variation in pressure along the centerline has the following shape:

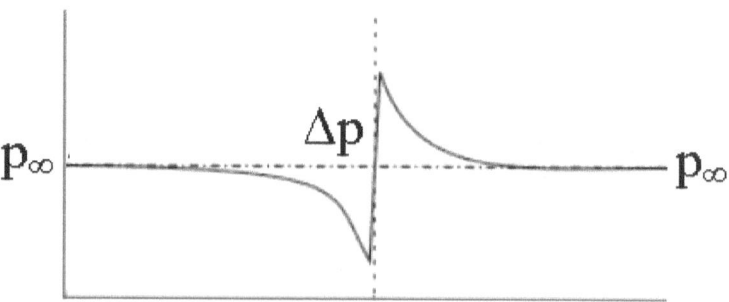

Figure 5. Pressure Variation along Centerline

The pressure drops as the flow approaches (is sucked into) the propeller and rises as it leaves (is pushed out of) the propeller. We now seek a relationship between the pressures and velocities.

2

Bernoulli's Equation

This important relationship covered in every introductory course in fluid flow dates back to the Egyptians and was formally published in 1738 by Bernoulli in his book, *Hydrodynamica.*[1]

The relationship is often stated:

$$\frac{V^2}{2g} + \frac{p}{\rho g} + h = cons \qquad (1.8)$$

Sometimes this is multiplied by density and/or the acceleration of gravity. In the form above, the units are length, often referred to as *head*; hence the symbol h. The following figure illustrates what the Egyptians figured out:

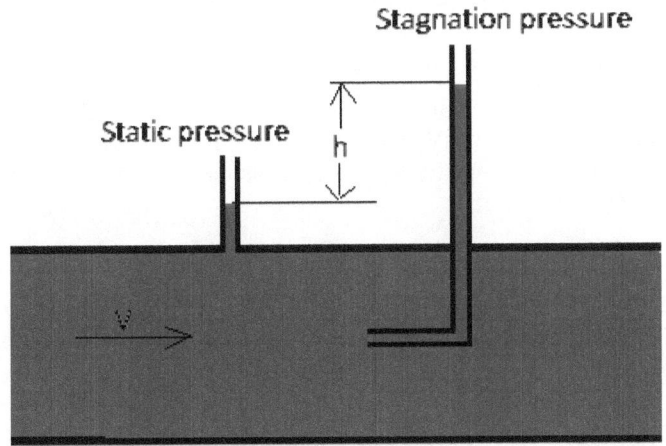

Figure 6. Static vs. Stagnation Pressure

The head is equal to $h=V^2/2g$. Because of the units, especially when multiplied through by ρg, it is often presumed (and sometimes incorrectly stated) that Bernoulli's Equation is an expression of energy, but this is not strictly true. Formally, Bernoulli's Equation is obtained by integrating the Navier-Stokes Equation[2,3] (i.e., the equation of fluid motion representing the conservation of linear momentum) along a streamline. Yes, integrating linear momentum over a distance yields energy, as in the definition of work:

$$W = \int \vec{F} \bullet d\vec{X} \qquad (1.9)$$

---

[1] Daniel Bernoulli (1700–1782) Swiss mathematician and physicist.
[2] Claude-Louis Navier (1785–1836) French engineer and physicist a pioneer in the field of mechanics.
[3] George Gabriel Stokes (1819–1903) Irish physicist and mathematician.

but viewing it in this way would obscure all of the assumptions that are made to arrive at Equation 1.8, namely: steady-state, constant density, incompressible, negligible viscous effects, constant gravity, as well as more subtle ones. Bernoulli's Equation is *only* valid along a streamline. It most assuredly does not apply across streamlines. It can also be reasonably accurate in some cases when energy is not constant from one point to the next, as with viscous dissipation. From Bernoulli's Equation, we see that the pressure difference across the propeller is equal to the difference in the square of the velocities before and after, as the elevations are the same for a horizontal centerline. This is an approximation, as it ignores the fact that the streamlines actually swirl around the propeller, which we will eventually discuss.

$$F = \frac{\pi D^2}{4}\left(V_4^2 - V_1^2\right) \tag{1.10}$$

Power and Efficiency

When attached to a moving plane or boat, the propeller is advancing through the water at velocity V_1, even though the fluid (air or water) is moving through the propeller at velocity V_2. This means that the rate of net work being accomplished is equal to:

$$\dot{W}_{net} = FV_1 = \frac{\pi D^2}{4}V_1\left(V_4^2 - V_1^2\right) \tag{1.11}$$

Ignoring swirl and viscous dissipation, the gross rate at which work is being supplied is equal to:

$$\dot{W}_{gross} = FV_2 = \frac{\pi D^2}{4}V_2\left(V_4^2 - V_1^2\right) \tag{1.12}$$

This means that the highest efficiency any propeller can possibly achieve is:

$$\eta_{max} = \frac{V_1}{V_2} \tag{1.13}$$

Swirl and viscous dissipation losses are in addition to the loss inherent in the delivery process. Taking the partial derivative of Equation 1.11 with respect to V_1 and setting this to zero gives us the point of optimum performance:

$$\frac{\partial P}{\partial V_1} = \frac{\pi D^2 \rho}{4}\left(V_4^2 - 3V_1^2\right) = 0 \tag{1.14}$$

This can be rearranged to yield:

$$\frac{V_4}{V_1} = \sqrt{3} \qquad (1.15)$$

Combining Equations 1.13, 1.14, and 1.15 yields:

$$\eta_{opt} = \frac{2}{1+\sqrt{3}} = 73.2\% \qquad (1.16)$$

The trivial maximum efficiency may be unity (ignoring swirl and viscous dissipation), occurring when $V_1=V_2$, but there is no thrust at this point, so this case is of no interest. What does this mean practically? Consider this: If you want to go 60 mph, you need to plan on pushing water (or air) out the back at 104 mph.

Pitch

The pitch of a fan or propeller is roughly defined as the forward advance resulting from a single rotation without slip. The speed of the fluid flowing through the propeller is approximately equal to the rotational speed times the pitch, φ. From trigonometry, we conclude that:

$$\varphi = \pi D \tan(\theta) \qquad (1.17)$$

The angle, θ, is rarely constant over the face of the propeller, so that we might calculate some average using a double integral:

$$\varphi = \frac{\iint \pi D \tan(\theta)\, dA}{A} \qquad (1.18)$$

Consider the previous goal of pushing water at 104 mph. For example, with a 1:2 gear ratio and engine rotation of 5500 rpm, this would require a pitch of 40 inches (slightly more than one meter).

$$pitch = \frac{(104mph)\left(5280\frac{ft}{mile}\right)\left(12\frac{in}{ft}\right)}{\left(\frac{5500rpm}{2}\right)\left(60\frac{min}{hr}\right)} = 40in \qquad (1.19)$$

You would have a difficult time purchasing such a prop for an outboard motor and it would require considerable power to turn it. From Equation 1.17 we see that for a twelve-inch (one foot) diameter propeller to have a forty-inch pitch, would require an average angle, θ, of 47°. Even with an angle of 45° we can expect considerable power to be dissipated as swirl. The prop in Figure 1 on page ii is quite unusual in that the pitch is equal to twice the diameter.

Chapter 2. Marine Propellers

The shape of a propeller reveals its intended application. Large sweeping blades are used when thrust is of primary importance and small slicing blades are for speed. Four or more blades are preferable for pulling and most racing props have only two blades. The two most important dimensions of a propeller are pitch and diameter. Pitch is the distance traveled in one revolution if there were no slippage. The diameter is larger than the pitch for thrusting props and the pitch is larger than the diameter for racing props. Figures 1 through 3 (pages ii and iv) are racing props; whereas, the one pictured on the front cover and below is for a tugboat:

Figure 7. Tugboat Prop

There are several differences between these two types of propellers, including: 1) the angle of the leading edge of the blade with respect to the shaft. The tugboat prop has a modest angle with a smooth progression; whereas, the racing prop has an aggressive angle with an abrupt ending. Blades of the racing prop are curled inward (with the flow and toward the shaft); whereas, blades of the tugboat prop are curled outward (against the flow and away from the shaft). The purpose of these two different curls is to produce either a convergence or divergence of the flow, as illustrated in the following figure:

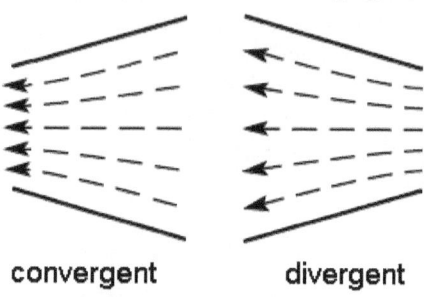

convergent   divergent

Figure 8. Converging vs. Diverging Flow Patterns

Converging flow produced by the inward curled blades of the racing prop tend to increase velocity (reduce area) and speed of forward movement; whereas, diverging flow produced by the outward curled blades of the tugboat prop tend to decrease velocity (increase area) and potential thrust. A speedboat could never dock a ship nor could a tugboat win a race.

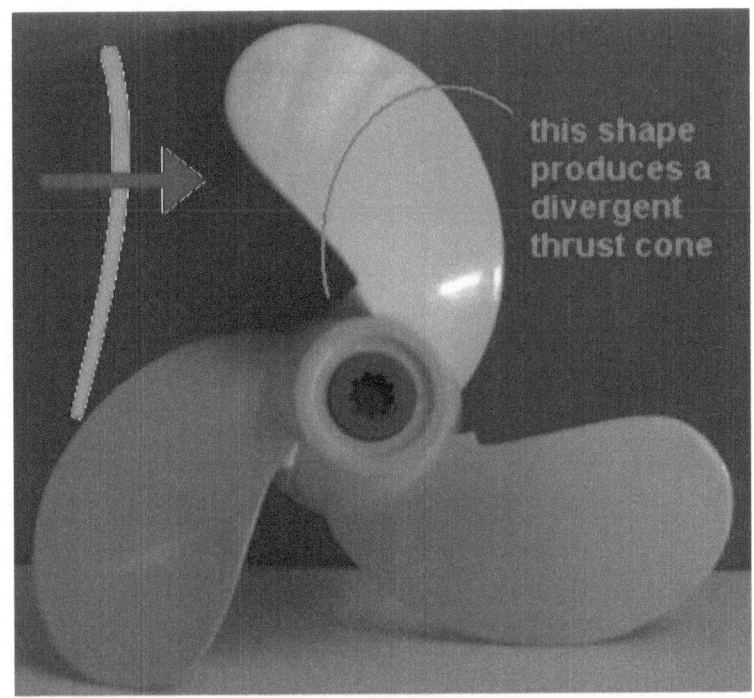

Figure 9. Shaped to Produce Divergent Thrust Cone

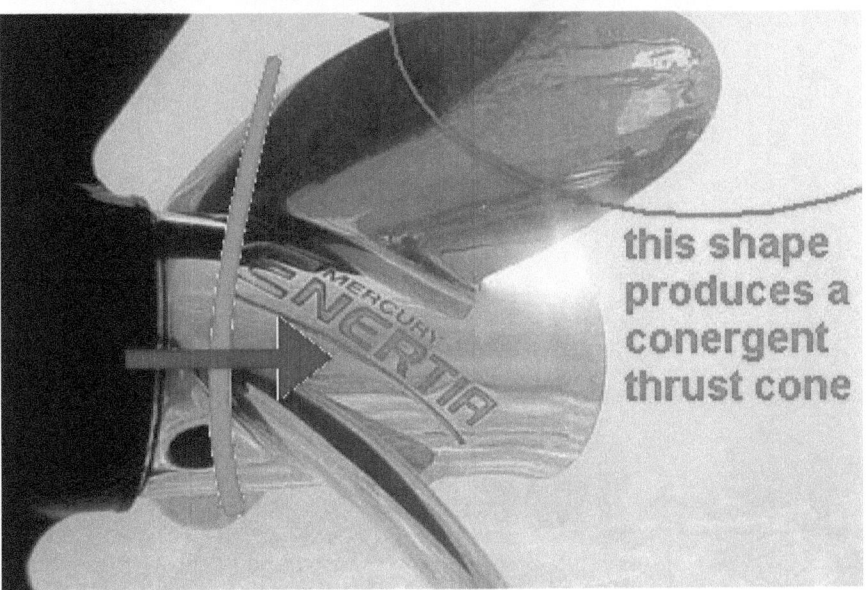

Figure 10. Shaped to Create Convergent Thrust Cone

9

Cavitation bubbles visualize the thrust cone in the following picture:

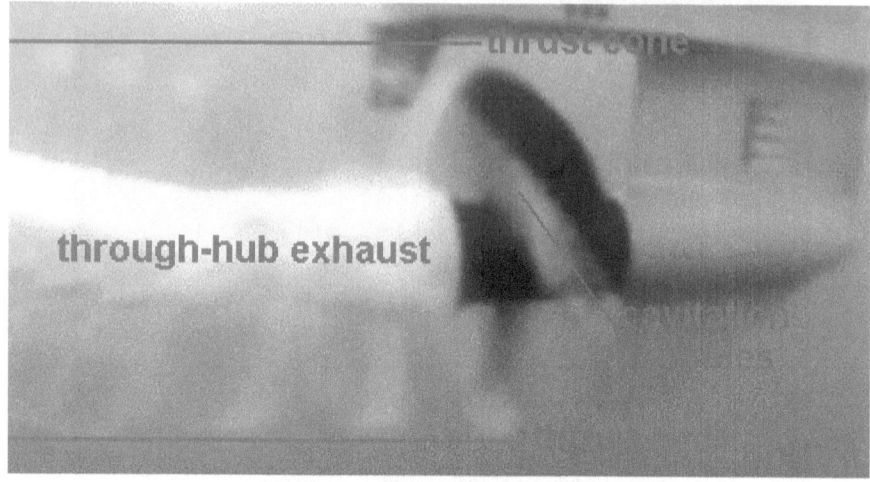

Figure 11. Balanced Thrust Cone

Notice that the cavitation bubbles seem to appear behind the prop. In this case, these are mostly forming along the trailing edge of the blades. Cavitation bubbles can also form on the leading edge of the blades, given different conditions.

Bubbles from cavitation and through-hub exhaust reveal how this thrust cone diverges over distance:

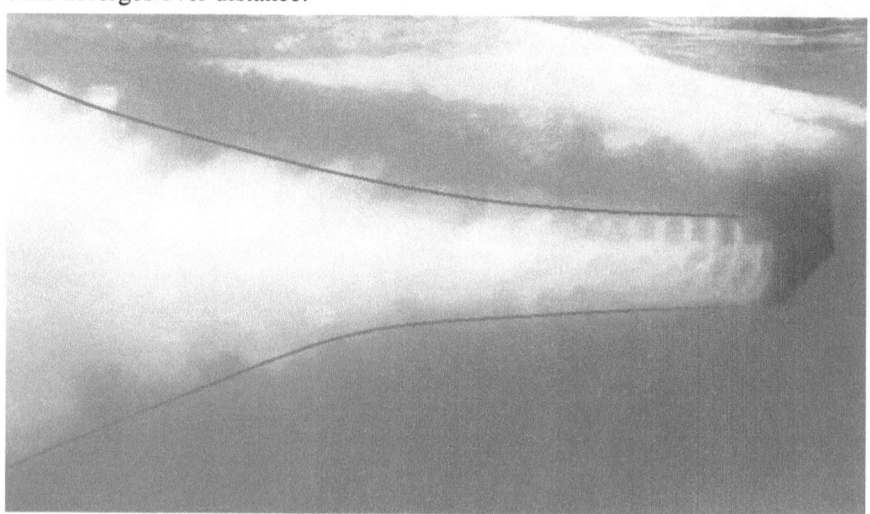

Figure 12. Bubbles Reveal Thrust Cone Shape

There are other objectives besides thrust and speed, such as running through weeds without being entangled, which is the purpose of the following prop. The backward sweeping blades tend to throw off weeds. The aggressive leading edge of the racing props shown in Figures 1 through 3 would tend to snag weeds and draw them inward, rather than pushing them outward.

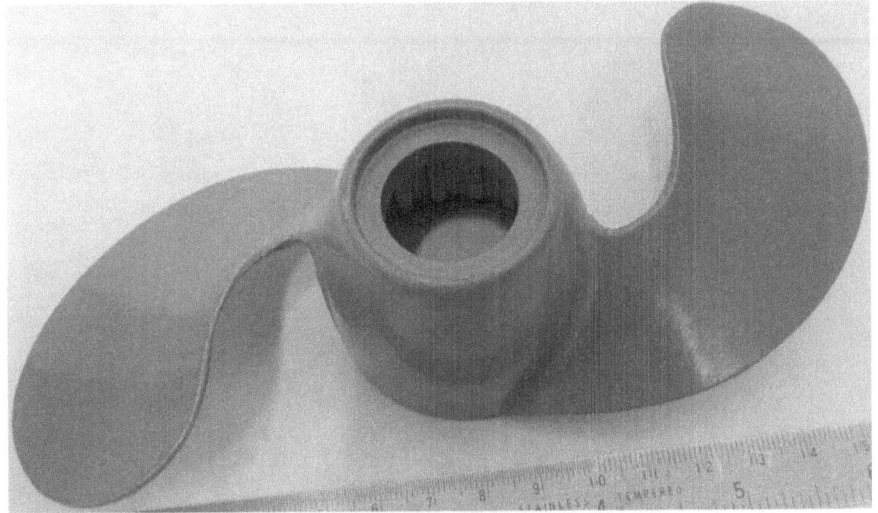

Figure 13. Weedless Prop

Some hull designs and engine placement result in stern squat. Boats designed for wakeboarding are intentionally made this way. Leaving a large wake requires power so that, unless you intend to do so, it is quite wasteful. When speed is of interest, getting the hull to rise out of the water is critical to achieve minimal drag. One way of lifting the stern is through propeller design through what is called a *cleaver*, which derives it's name from the type of blade used to chop meat. In this design, the trailing edge of the blades ends abruptly, as illustrated in the next figure:

Figure 14. Cleaver Design with Truncated Trailing Edge

The increased stern lift of this design comes at a price, which is less control over the thrust cone, often called *hole shot*. The many differing objectives give rise to diverse prop designs. One creative idea is to mount two contra-rotating propellers on the same shaft. The concept behind this is an attempt to reduce the large-scale swirling motion created by a single propeller. The effectiveness of this design at reducing swirling vortices can readily be seen in the near wake even at low speeds.

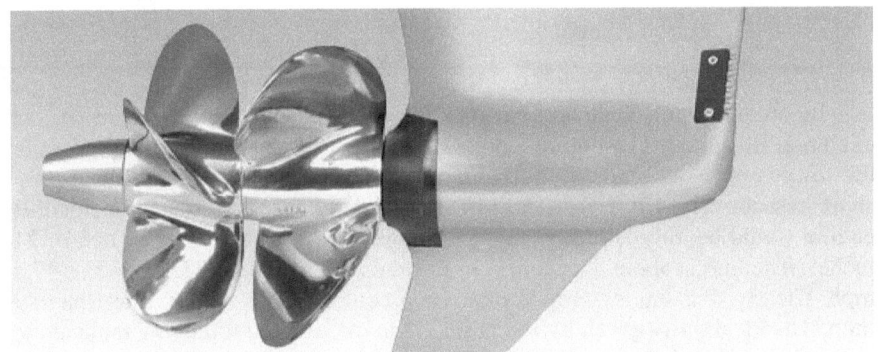

Figure 15. Contra-Rotating Props on a Single Shaft

Only the increase in velocity along the direction of motion is useful. Swirl and turbulence are just wasted power. The design above is demonstrably more efficient than the arrangements in Figures 11 or 12. As the old saying goes…

You don't get something for nothing.

Think about it… Two props turning in opposite directions close to each other will have twice the difference in velocity of one prop rotating relative to a strut or gearbox and in as close or even closer proximity. This means far greater *shear*. Shearing motion is inherently opposed by viscosity in all fluids. Newtonian shear stress is equal to the viscosity times the rate of change in velocity. Two contra-rotating props exchange large-scale swirling motion for small-scale shearing motion. Overall it may be somewhat more efficient, but not without consequence—the very least of which is depression of the thrust cone, which is adverse to achieving high speed. You might want to utilize this design on a party boat or transport aircraft, but not on a race boat or fighter plane.

<u>More Detailed Calculations</u>

After considering some actual shapes, including the shaft (or hub), and recalling the definitions of V_1 through V_4 from Figure 4, we can improve upon Equation 1.11. We must consider, not only the diameter of the prop, D_P, but also the diameter of the hub, D_H. Here we also substitute the increase in fluid velocity affected by the prop, ΔV, relative to the forward motion, V. A more accurate expression for the thrust would then be:

$$F = \frac{\pi\left(D_P^2 - D_H^2\right)}{4} \rho \Delta V \left(V + \Delta V\right) \qquad (2.1)$$

The first group in this equation is the effective area, ρ is the density, V is the velocity of the incoming flow, and ΔV is the increase in velocity provided by the propeller. If there were no slip, the average velocity of the fluid at the propeller would be, V_P:

$$V_P = \Omega\varphi \qquad\qquad (2.2)$$

where Ω is the rotational speed, and φ is the pitch.

In order to put some numbers to these equations, we now consider an example. Imagine a 115 hp outboard motor turning 5800 rpm. Typical gears in the lower unit would have a tooth count of 12 and 28 for the pinion and prop shaft, respectively, for a gear ratio of 2.33:1. The most common prop for this engine would be a three-blade having a diameter of 13 inches and a pitch of 21 inches. The hub is about 4⅛ inches. In this case V_P would be 72.5 ft/sec or 49.4 mph. Clearly, this engine/prop combination couldn't push a boat faster than 49 mph. The ideal thrust given by Equation 2.1 is listed in the following table along with other calculations:

	A	B	C	D	E	F	G	H	I	J	K
1	typical 115 hp outboard										
2	5800	rpm									
3	2.33	gear ratio									
4	13	diameter [in]									
5	4.16	hub [in]									
6	21	pitch [in]									
7	72.5	Vp [ft/sec]									
8	62.4	density [lb/ft³]									
9	boat	boat	thrust	gross	prop	prop	prop	thrust	gross	net	prop
10	mph	ft/sec	lbf	hp	slip	ft/s	mph	lbf	hp	hp	eff
11	0.0	0	8433	**1112**	68%	23.3	15.9	872	115	0	0%
12	3.4	5	7852	1035	64%	26.0	17.7	872	115	8	7%
13	6.8	10	7270	958	60%	28.8	19.7	872	115	16	14%
14	10.2	15	6689	882	56%	32.0	21.8	872	115	24	21%
15	13.6	20	6107	805	51%	35.4	24.1	872	115	32	28%
16	17.0	25	5525	728	46%	39.0	26.6	872	115	40	34%
17	20.5	30	4944	652	41%	42.7	29.1	872	115	48	41%
18	23.9	35	4362	575	36%	46.7	31.8	872	115	56	48%
19	27.3	40	3780	498	30%	50.7	34.6	872	115	63	55%
20	30.7	45	3199	422	24%	54.9	37.4	872	115	71	62%
21	34.1	50	2617	345	18%	59.2	40.4	872	115	79	69%
22	37.5	55	2036	268	12%	63.6	43.3	872	115	87	76%
23	40.9	60	1454	192	6%	68.0	46.4	872	115	95	83%
24	**42.5**	62.33	1183	156	3%	70.1	**47.8**	**872**	**115**	**99**	**86%**
25	44.3	65	872	115							
26	47.7	70	291	38							
27	49.4	72.5	0	0							

Figure 16. Example Calculations for 115 hp Outboard

The gross power is equal to the thrust times the velocity at which it is delivered, or the prop speed. Clearly, this engine can't produce 8433 pounds of

14

thrust and 1112 hp. Recall that Equation 2.1 assumed there was no prop slip. If we include enough slip so that the gross power delivered is equal to 115 hp, we get columns E through K.

The net power is equal to the thrust times the forward velocity, or the boat speed. The boat on which this example is based tops out at 42.5 mph. If the engine develops 115 hp, then the thrust would be 872 pounds and the net would be 99 hp. At top speed the prop slip would only be 3% and the prop efficiency would be 86%, as shown in the table. This next figure shows the prop speed and slip over the range of boat speed (at full-throttle, of course):

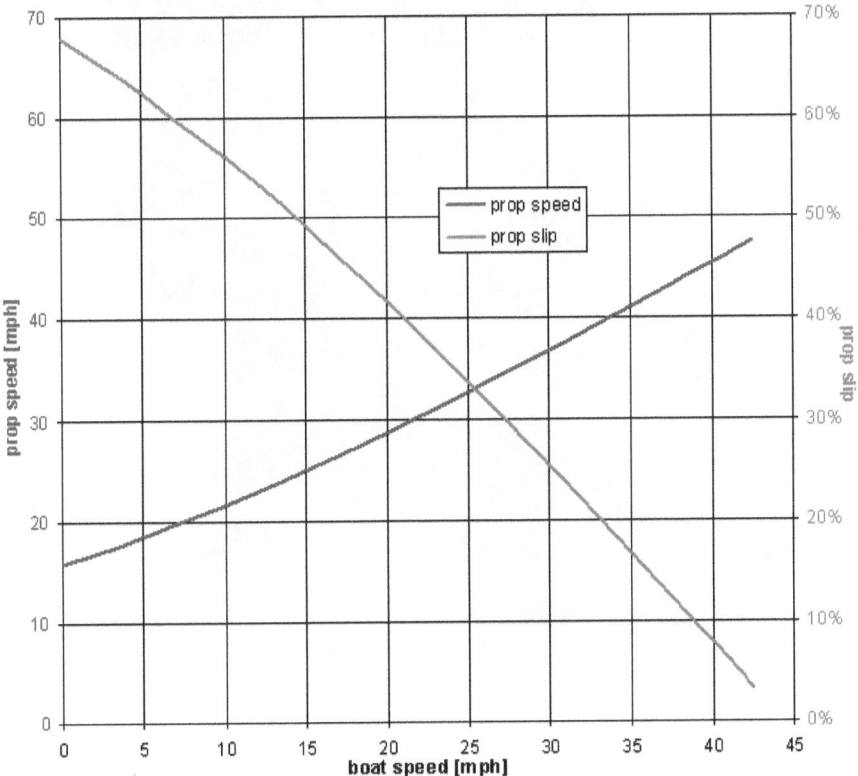

Figure 17. Calculated Prop Speed and Slip

The thrust is proportional to the area times the velocity times the change in velocity and the power is equal to the thrust times the velocity, so the power required to turn a propeller is proportional to **length^5** times **rate of rotation^3**. The power is also proportional to the density, which is why airplane propellers are so much larger than marine props.

$$P \propto \rho D^5 \Omega^3 \tag{2.3}$$

15

If you want to go fast, you must throw water (or air) even faster. As the engine can only deliver so much power, you must throw a lot less water faster, which means a much smaller, faster rotating prop and a 1:1 gear ratio. This next figure illustrates this relationship of scale:

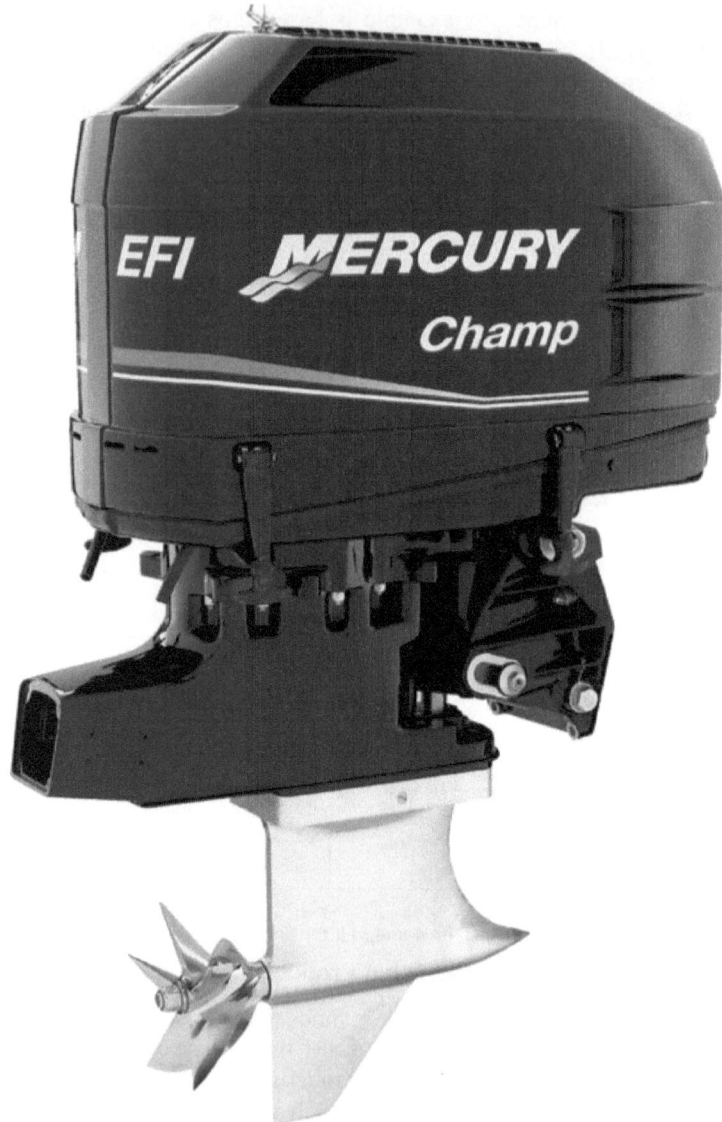

Figure 18. BIG Engine with Very Small Prop

Modifications

Various modifications have been developed to enhance certain aspects of marine propellers. Avoidance and control of cavitation has been the motivation for much of this effort. Cavitation not only robs power, it causes considerable damage, as illustrated in this next figure:

Photo of a blade damaged by Cavitation Erosion

Figure 19. Cavitation Damage

The most common modification to control cavitation is called *cupping*. Traditional cupping involved slightly rolling up the trailing edge, as illustrated in the following figure:

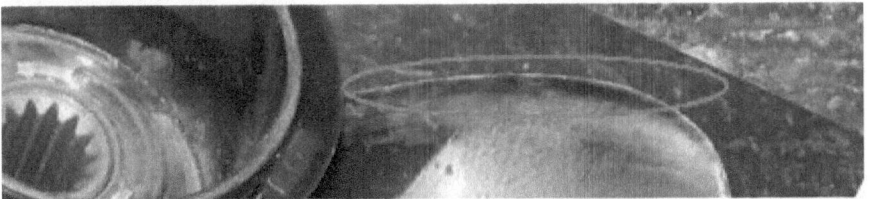

Figure 20. Slightly Cupped Trailing Edge

The purpose of this rolled edge is similar to the wing tip fins that have become quite common on commercial aircraft. These were originally added to large planes to discourage vortices from rolling off the end of the wing and possibly upsetting the flight of smaller aircraft.

Figure 21. Aircraft Trailing Vortices

Once the airlines discovered that these wingtips slightly improved performance, they started putting them on planes of all sizes.

Figure 22. Upturned Wingtip Reduces Lateral Vortex Shedding

"If a little is good, then a lot must be better," has always been a tendency with me. I applied this reasoning to cupping props in 1966.

Figure 23. Aggressive Trailing Edge Cupping

The rolled edge is much deeper and extends all the way around the end of the blade.

Figure 24. Cupping Locations on Trailing Blade Edges

There are better ways of accomplishing this, but I used a trailer hitch ball and a hammer. I shared this secret with a friend, Bruce Borkenhagen. He won an outboard regatta championship and the other racers had to know why. Bruce eventually went to work for OMC racing and testing engines. I'm glad the secret got out. Someone has taken this approach even further.

Figure 25. Piranha-Style Cup

Slip

Our definition of pitch just before Equation 1.17 included the caveat: *without slip*. Propellers operating in any fluid will have some amount of slip; that is, the fluid being forced through the prop will on average be moving at a speed less than $\Omega\varphi$. Just how much slip is quite difficult to measure but anecdotal information does exist plus we know the limits. If the prop were advancing through the fluid at $\Omega\varphi$, there would be no thrust ($\Delta V=0$) and no force to induce slip. A boat could not achieve this unless it were being towed or there were more than one engine. An airplane can achieve this in a dive. As we will see in the next chapter, fans can achieve this in a laboratory when there is another fan in the same duct to provide the flow.

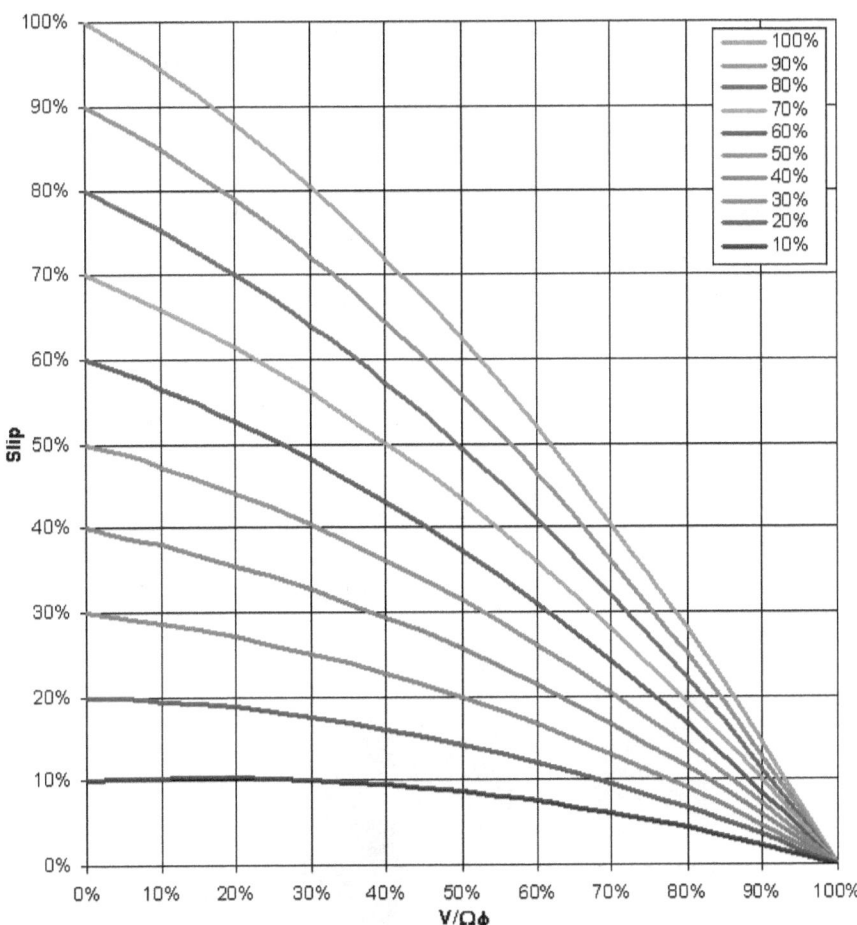

Figure 26. Slip vs. Speed Ratio

All of the curves must go through zero at $\Omega\varphi=1$, which defines the right side. The left side is the slip with no forward movement (i.e., when the boat is tied to the dock or a tugboat is struggling to move a big ship or barge). The left side is roughly equal to the *open* area of the prop. A complete *screw* would have no open area and very little slip, regardless. The speed prop in Figure 1 has a rather high open area, about 65%. The tugboat prop in Figure 7 has a much smaller open area, about 27%. Again, the propeller design is driven by its intended use. In the previous example of the typical 115 hp outboard, we assumed an open area of about 50%.

As the angle of attack, θ, approaches 90°, there is only swirl and no longitudinal motion or thrust along the shaft. From Equation 1.17 we see that the ratio of the pitch to the pitch plus diameter is indicative of this effect and also dimensionless. We assign the symbol Ψ to the slip at zero $V/\Omega\varphi$ and the symbol Γ to the open area fraction. From the limiting case, we conclude that:

$$\Psi = 1 - (1-\Gamma)\left(\frac{\varphi}{\varphi+D}\right) \tag{2.4}$$

The slip is then equal to:

$$S = \left(\Psi + \frac{cV}{\Omega\varphi}\right)\left(1 - \frac{V}{\Omega\varphi}\right) \tag{2.5}$$

where c is a geometric constant. Given a relationship for hull drag and these equations, we might evaluate various prop options using an Excel spreadsheet.

Figure 27. Slight Cup on Five-Blade SS Prop

Hull drag depends on many factors, including: speed, geometry, angle, and loading. These topics are beyond the scope of this text. We can, however, select some available curve and adapt it for our purposes here, as with the next figure.

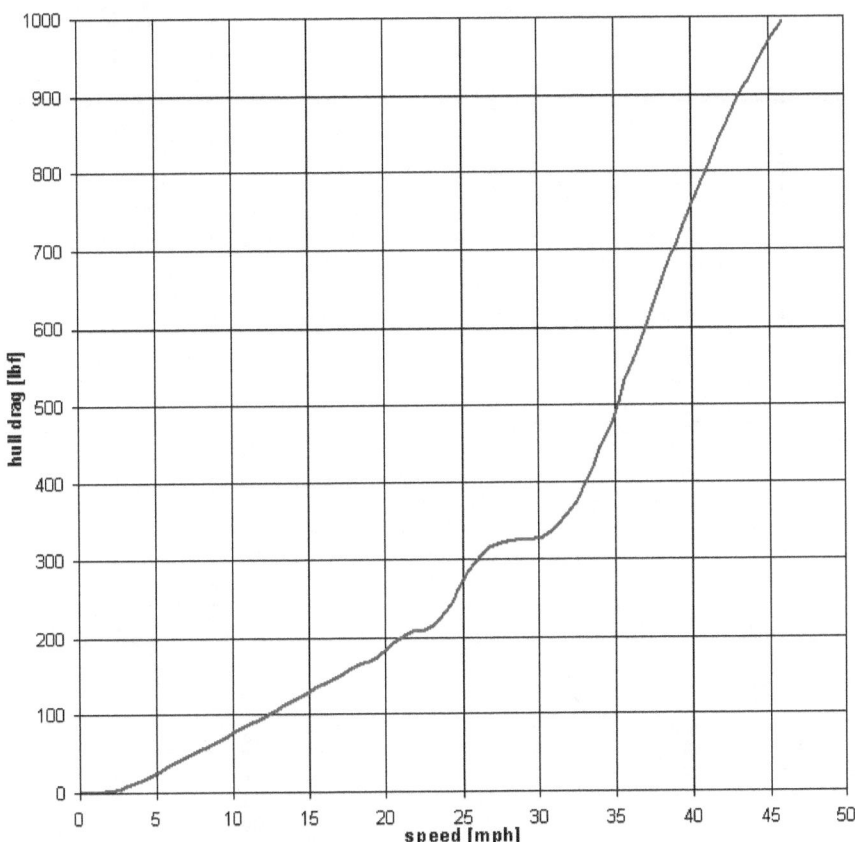

Figure 28. Example Hull Drag vs. Speed

We would also need a curve of torque vs. rpm for the engine, as prop selection will definitely impact rpm and the torque is not constant.

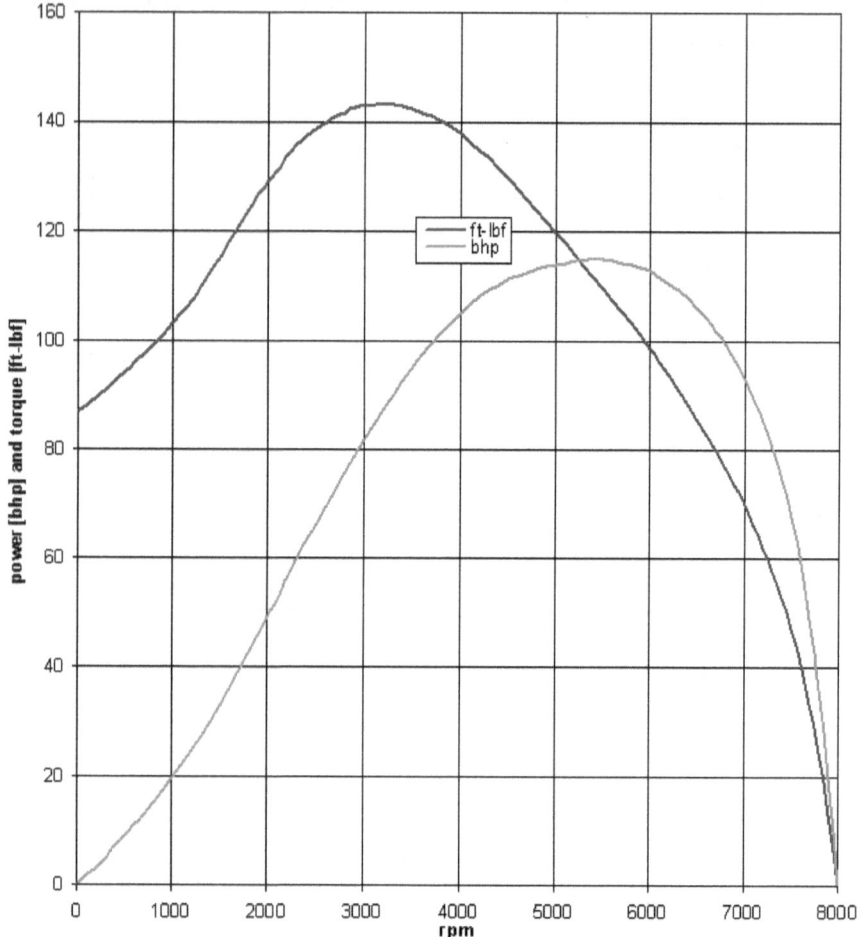

Figure 29. Torque and Power Curve for Two-Stroke Engine

We can combine these calculations and curves into a prop selection spreadsheet (boat.xls), which you will find in the online archive in the examples folder. In this we step through a series of prop diameters (10" to 15") and pitches (16" to 32") to arrive at the following results (see examples\boat.xls):

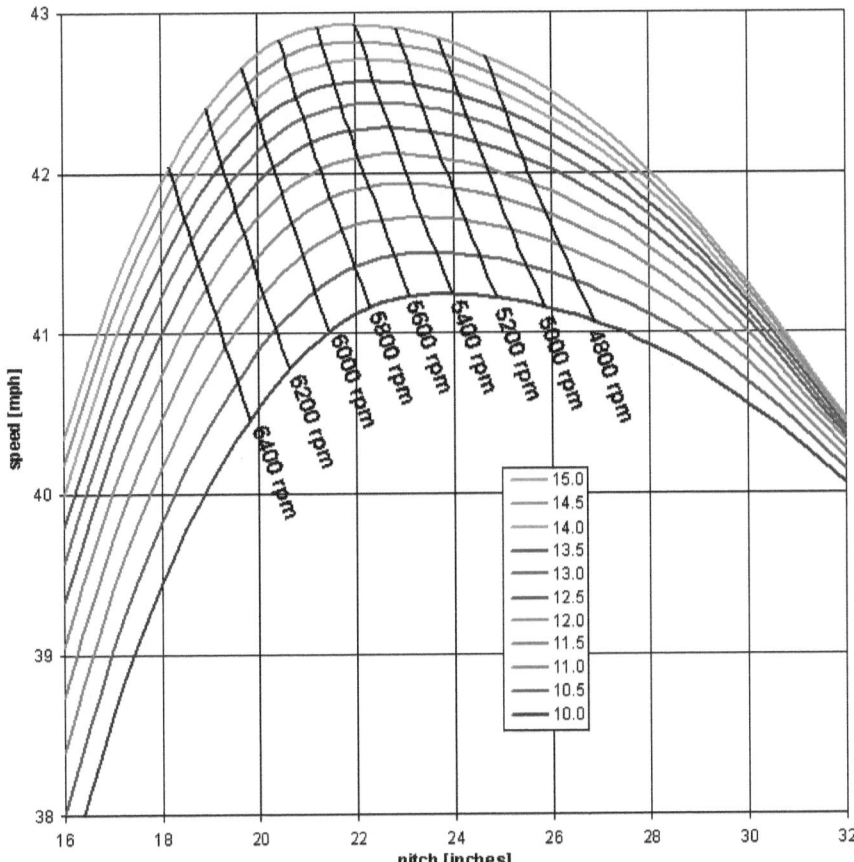

Figure 30. Hull Speed vs. Prop Pitch and Diameter

Not surprisingly, the peak (optimal pitch) changes with diameter. The engine rotational speed is an integral part of the calculation, which is performed using Excel's Solver to match thrust and drag plus power required and delivered. Of course, a smaller prop can be turned more rapidly (higher engine rpm) so that the engine rotational speed varies with both diameter and pitch. While this engine may be capable of turning 7000 rpm, it won't last long at that rate and it is unreasonable to select a prop resulting in such high rpm. We must, therefore, restrict our selection to those combinations of diameter and pitch.

25

Chapter 3. Axial Fans

In this chapter we will consider only large shrouded axial fans moving air. The following figure shows a typical fan atop a cooling tower along with three men to convey the scale. Notice the shape of the blades.

Figure 31. Typical Cooling Tower Fan

Such fans are often 30 feet in diameter and are driven by electric motors developing 200 hp or more. Performance of these fans depends on many factors, including: size, design, number of blades, and pitch. While the blades of marine props often have uniform thickness, axial fan blades are often airfoil-shaped:

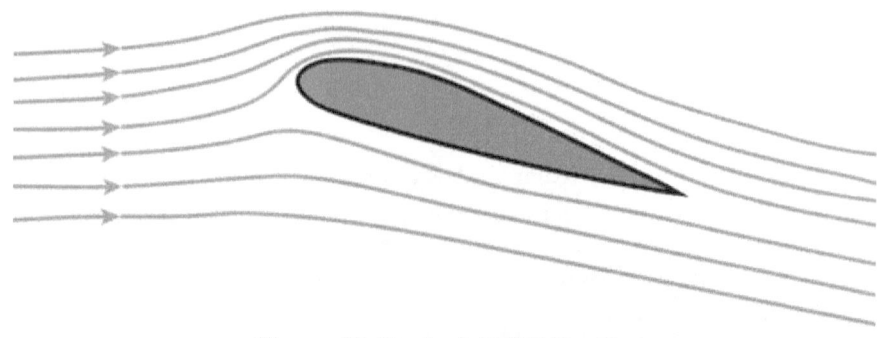

Figure 32. Typical Airfoil Profile

While these fans can be tested in place while operating, performance curves are often based on laboratory measurements. Fan test facilities are a similar to wind tunnels.

Figure 33. Fan Test Tunnel

The pressure rise across the fan, volumetric flow rate, torque, and power are measured over a variety of conditions, including rotational speed. The flow can be controlled by louvers, such as are shown at the right end of the tunnel above.

Static vs. Total Pressure

Before we delve any further into this topic, it is essential to dispel a great error common in the literature surrounding fans. The *static* (i.e., stationary, non-moving) pressure is simply that—the pressure irrespective of the fluid motion. The *total* (i.e., moving, combined, stagnation, augmented, apparent, etc.)

28

pressure is equal to the static pressure plus the velocity head ($\rho V^2/2$). Recall this term from Bernoulli's Equation in Chapter 1.

It can be extremely difficult (i.e., problematic), even virtually impossible, to *measure* the static pressure in a complex flowing environment, such as that surrounding a fan. Not that instruments (i.e., gauges or manometers) are unavailable or inaccurate. The problem is not only *how* but *where* would you measure the static pressure? If you stick a probe out into the flow, you will not be measuring the static pressure, regardless of how you position the probe or where you locate the sampling ports.

Consider the following common probe often used for measuring water flow:

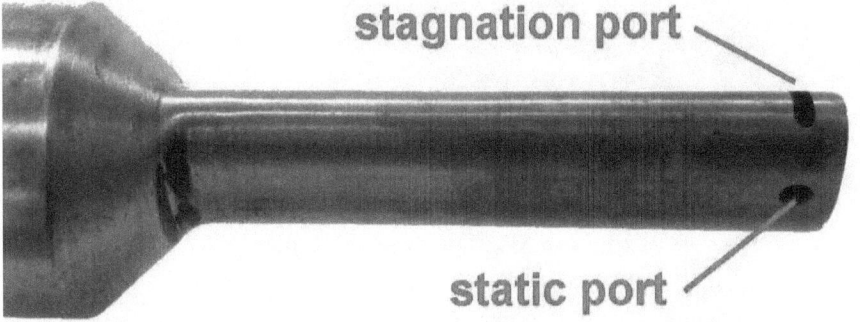

Figure 34. Pitot Probe

The stagnation port is directed into the flow and the static ports (one on either side) are perpendicular. Consider the classic case of flow over a cylinder for which much experimental data has been collected:

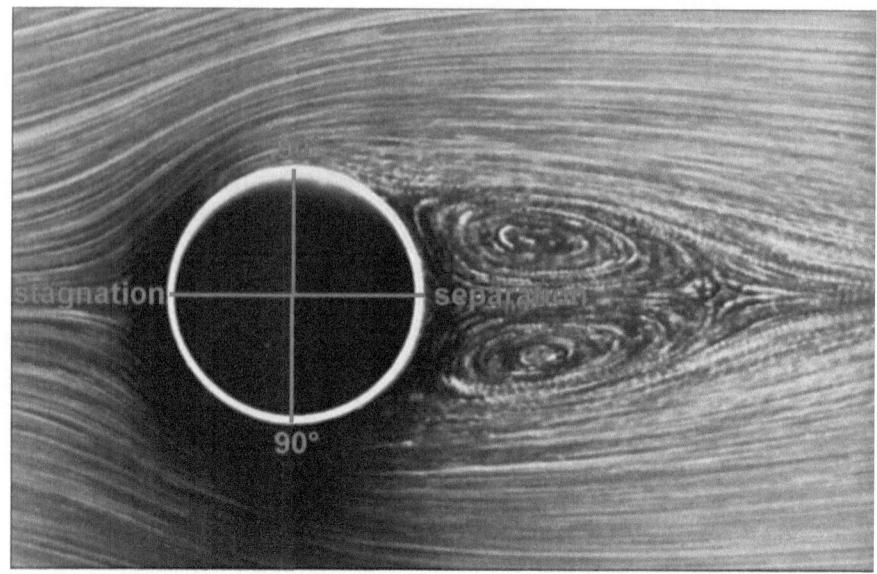

Figure 35. Flow over a Cylinder

The pressure at the leading edge is most likely equal to $P_{STATIC}+\rho V^2/2$, the stagnation pressure. The pressure on the back side in the separation zone may be equal to something near $P_{STATIC}-\rho V^2/2$. We *hope* the pressure at the static ports (located approximately 90° on the side of the probe) will be close to P_{STATIC} so that the difference will be $\rho V^2/2$. There is no physical reason why the pressure at the side ports *must* be *exactly* equal to the static pressure and there is ample experimental evidence that this is not always the case. Hence, measuring the static pressure in a flow is problematic to say the least.

Out of the main flow, off in a corner, behind a bolt or flange, along a piece of angle iron, next to a column, or beside a motor mount, you just might measure the static pressure with an instrument. The easiest thing to do experimentally is connect one end of a pressure sensor or manometer to the stagnation port of such a probe and the other end to a convenient place out of the flow on the upstream side of the fan. You then systematically pass the probe across the flow downstream of the fan. What you're measuring will be the downstream total pressure (static plus velocity head) minus the upstream static. This is most assuredly NOT the pressure difference across the fan! It is the pressure difference across the fan PLUS the velocity head.

Figure 36. Cement Block Wall Fitted with a Fan

Two points are crucial here: 1) the pressure difference in Equation 1.7 is STATIC, not static plus the velocity head and so with the thrust in Equation 2.1; and 2) the P term in Bernoulli's equation is STATIC pressure. And that's not all... Recall one of the caveats associated with Bernoulli's Equation? It is NOT valid ACROSS a streamline, only ALONG a streamline. Air does not flow in smooth lines from a corner behind a flange in the back of the plenum and out through the center of the fan. Bernoulli's Equation is not valid along that imaginary path.

The only thing that matters when it comes to the net work provided by a fan is the difference in STATIC pressure. This times the area is equal to the lateral force on the fan, were you to measure that in a test facility with force or strain gauges. If you were to construct a freestanding cement block wall, leave a hole, and put a fan in it, as illustrated in the preceding figure, lateral force on the wall would equal the difference in static pressure across the fan times the area.

Bernoulli's Equation does apply along a streamline approaching the upstream side of the fan from far away and also from the downstream side of the fan to far away on the other side of the wall. Bernoulli's equation tells us that the static pressure on the upstream side of the fan will be $P_{AMBIENT} - \rho V^2/2$ and the static pressure on the downstream side of the fan will be $P_{AMBIENT}+\rho V^2/2$, making the difference in static pressure across the fan equal to ρV^2.

As the wall isn't moving and other than the possibility that you might want to create a breeze in the absence of a natural one, all of the power going into the fan mounted in the block wall will be dissipated as heat into the environment. There will be no net work performed by this apparatus. You will definitely pay (at least figurative, if not literally) to move the air from one side of the wall to the other. Though pointless, you might try to measure the pressure rise across the fan (Figure 5) and even calculate efficiency, but you don't get to add a velocity head ($\rho V^2/2$) to the pressure difference to make your fan appear more efficient that it actually is.

Just because it's easier to measure downstream total minus upstream static doesn't mean that you should ever report the performance of a fan based on this misnomer. You *must* subtract off the velocity head before reporting the efficiency. Misunderstanding of this principle is why reports exist in the literature of both *static* and *total* fan efficiency. Total efficiency sounds like it should be appropriate, but it is entirely bogus. The only quantity that matters for a fan is *static* efficiency. It should not even be necessary to clarify.

How can you tell? Recall when the forward motion of the prop is equal to the velocity of the fluid through the prop, there is no thrust and no net work being performed. At that point the efficiency is zero! When plotted vs. flow, the efficiency of a fan should rise to a peak and then fall back to zero. If it approaches some asymptote, rather than returning to zero as flow continues to increase, it's the wrong curve!

Failure to comprehend the distinction between static and total fan efficiency cost a fine company millions of dollars in lawsuits and sullied the reputation of more than one smart, honest engineer. I hope this never happens again, which is why I continue to harp on the subject. This is a painful reminder of how important it is to understand the physics underlying a device such as a fan.

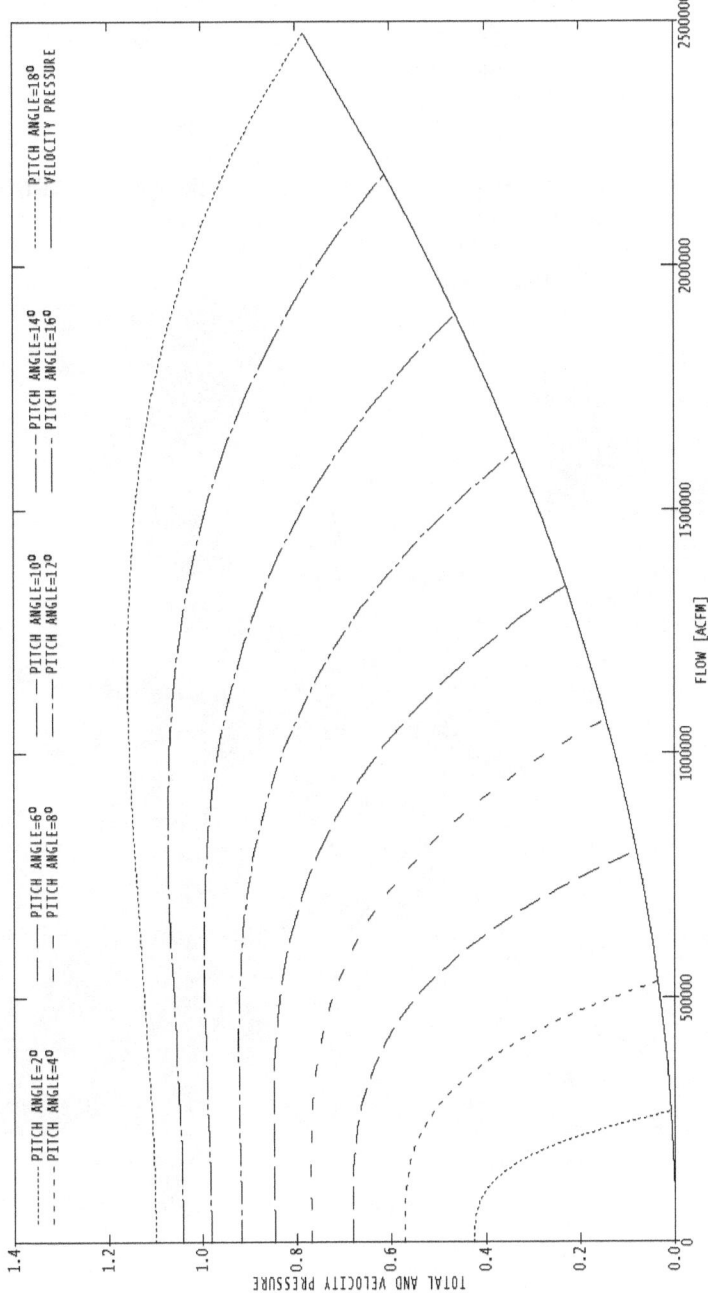

Figure 37. Confusing Fan Curves

33

The preceding figure is the shameless culprit in this sad story. The vertical axis is labeled "Total *and* Velocity Pressure." This is easily confused to imply "total *plus* velocity pressure", when what it actually means is "total and *also* velocity pressure." The velocity pressure (which we couldn't care less about here) is the upward curved solid line. The dotted and dashed curves of constant pitch angle all bend downward and stop where they intersect the velocity pressure curve. This is the point of zero delta-P (static pressure difference, of course, which is the measure we care about). The only way to reach this point in a test facility is with a second fan, which is *not* going to be in *your* cooling tower and so this has *no bearing whatsoever* on how *your* fan is going to operate.

Figure 38. There Is only ONE Fan in this Shroud

The way this information *should* be presented is shown in the following figure on the next page:

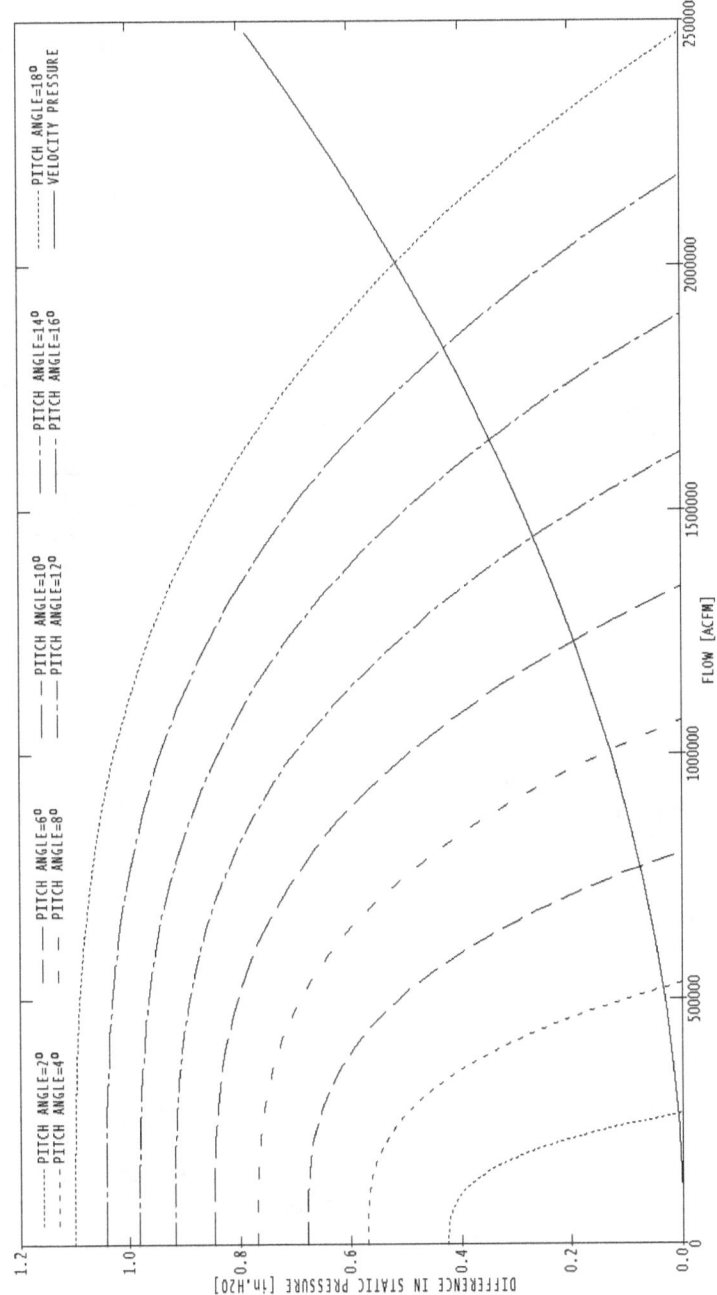

Figure 39. Unambiguous Fan Curves

35

Fan Test Data

The following is a typical set of efficiency curves for a cooling tower fan based on experimental data points:

	A	B	C	D	E	F	G	H	I	J	K	L	M	
		2"pitch		6"pitch		10"pitch		14"pitch		18"pitch		22"pitch		
1														
2	KCFM	Es	Et	Es	Et	Es	Et	Es	Et	Es	Et	Es	Et	
3	500	69.0%	74.5%	66.4%	69.9%	62.9%	65.4%	48.5%	50.3%	37.0%	38.3%	29.4%	30.4%	
4	600													
5	700													
6	800													
7	840													
8	900													
9	1000													
10	1070													
11	1100													
12	1200													
13	1300													
14	1310													
15	1400													
16														
17														
18														
19														
20														
21														
22														
23														
24														
25														
26														
27														

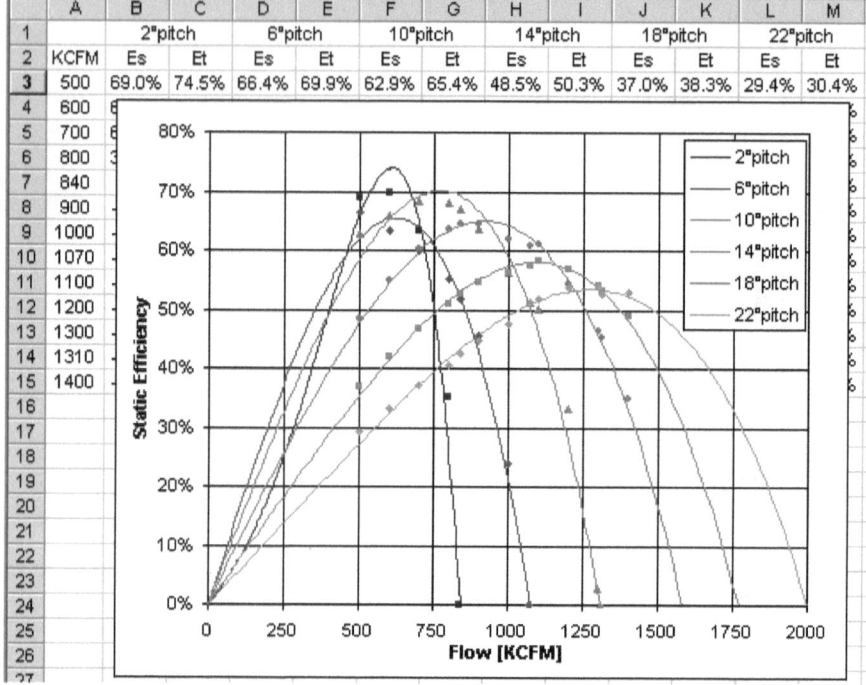

Figure 40. Typical Cooling Tower Fan Curves and Test Data

The curves are the manufacturer's advertised performance and the points are test data (i.e., measured results). For this type of fan the pitch angle is most often reported in degrees. This angle is in reference to an indicator bolted to each blade and is only approximate. The actual angle changes over the length of the blades.

These curves rise to a peak and then fall back to zero, as the velocity through the fan approaches $\Omega\varphi$, which we might term the *stall* speed. At this point, there is no thrust (line an airplane wing providing no lift), so the air wouldn't be flowing through the test facility if there weren't a second fan to push it. You can only reach this point with two fans, just like you could only reach such a point with a boat if it were being towed by another. The data, calculations, and graphs can be found in the online archive in spreadsheet examples\fan.xls. The information in the preceding two graphs are there as well.

36

If we divide the X-axis by the stall speed ($\Omega\varphi$) times the area ($\pi D^2/4$) and the Y-axis by the maximum value, the data collapse to a single curve:

	A	B	C	D	E	F	G	H	I	J	K	L
1	2"pitch		6"pitch		10"pitch		14"pitch		18"pitch		22"pitch	
2	V/Vs	Es	V/Vs	Es	V/Vs	Es	V/Vs	Es	V/Vs	Es	V/Vs	Es
3	59.4%	98.6%	46.5%	96.3%	38.2%	92.1%	31.5%	75.3%	28.2%	63.5%	25.0%	54.6%
4												81.7%
5												89.3%
6												75.5%
7												78.8%
8												83.1%
9												88.5%
10												95.1%
11												96.5%
12												99.8%
13												99.8%
14												98.0%
15												98.5%
16												94.2%
17												85.9%
18												83.6%
19												87.1%
20												52.6%

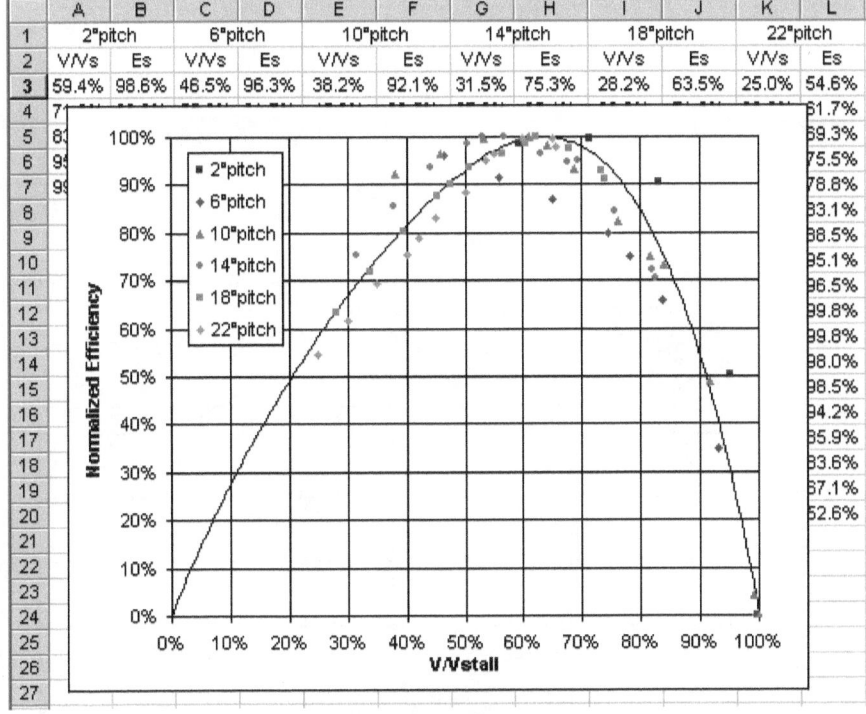

Figure 41. Fan Data Plotted on Transform Coordinates

The above curve for normalized efficiency vs. $\beta=V/V_{stall}$ is given by:

$$\frac{\eta}{\eta_{max}} = \frac{3}{4}\beta(1-\beta)((9\beta-3)\beta+4) \tag{3.1}$$

The maximum efficiency for an axial fan is limited by the swirl, as this consumes power, but doesn't produce thrust. This can be approximated by the following integral:

$$\eta_{swirl} = \int_0^1 \cos(\arctan(\varphi/x))\,dx \tag{3.2}$$

In the preceding equation $\varphi=\Omega/\pi D$ is the pitch ratio. This integral can be computed numerically and approximated by the following expression:

$$\eta_{swirl} = e^{\left[\frac{-1.17583\varphi}{1+(0.324221-0.00320923\varphi)\varphi}\right]} \tag{3.3}$$

37

The maximum pressure ratio, $\gamma_{max}=\Delta P_{max}/(\rho V^2/2)$, for an axial fan can be estimated by the following equation, which is based on a variety of test data for this type of fan.

$$\gamma_{max} = \frac{0.23965}{\varphi^{1.57314}} \tag{3.4}$$

The normalized pressure ratio can be computed with this next equation:

$$\frac{P}{P_{max}} = 1 - \varphi^3 \tag{3.5}$$

The static pressure difference developed by the fan pictured and the motor power required to turn it are shown in this next figure:

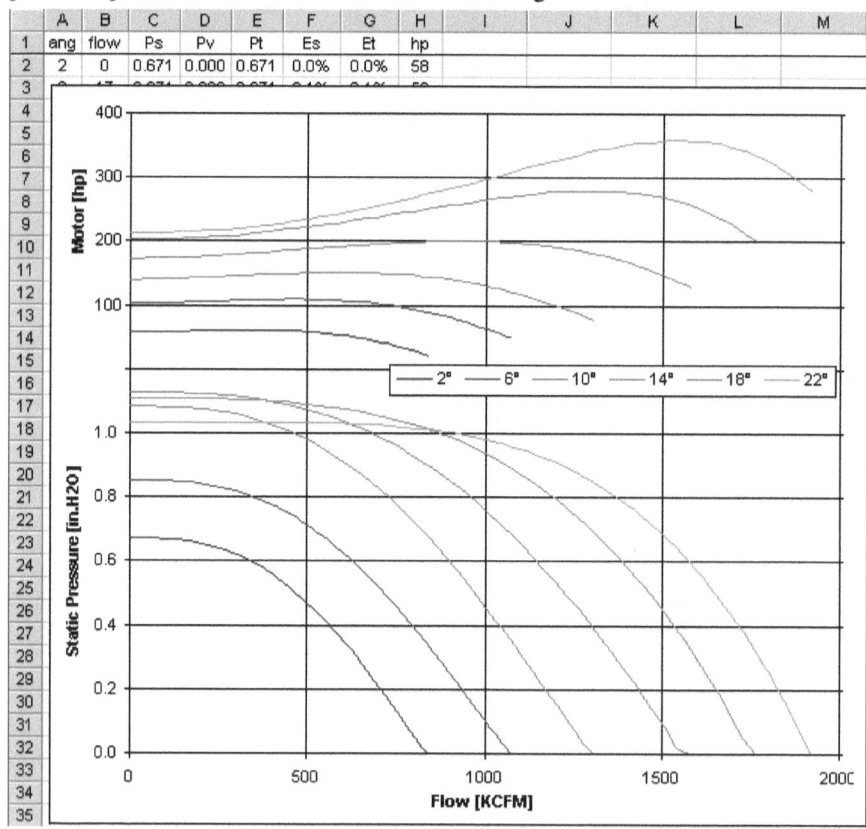

	A	B	C	D	E	F	G	H	I	J	K	L	M
1	ang	flow	Ps	Pv	Pt	Es	Et	hp					
2	2	0	0.671	0.000	0.671	0.0%	0.0%	58					

Figure 42. Theoretical Fan Curves

These curves were calculated using Equations 3.1 through 3.5 and are in excellent agreement with the manufacturer's published performance curves. The

38

static pressure curves are at the bottom and are read off the left scale. The power curves are at the top and are read off the right scale. This curve format is typical for the industry.

Innovative and Unusual Fan Designs

Just as propeller designs vary considerably, so do axial fan designs. Howden makes the most interesting blade shapes. This figure is from their online brochure.

Figure 43. Howden Fan

These fans are noticeably quieter than the common variety shown in the previous figures. They may have some performance advantage at certain operating conditions as well.

Sizing a Fan and Motor

Much confusion surrounds this rather simple topic. This confusion arises from a lack of understanding of basic fluid mechanics, for example, applying Bernoulli's Equation anywhere but along a streamline or adding extraneous quantities, such as velocity heads, for convenience or to embellish a plot. We need look no farther than the nearest textbook or handbook for guidance regarding these calculations. Turn to the chapter on pipe flow calculations and you will find the following figure for entrances and exits.

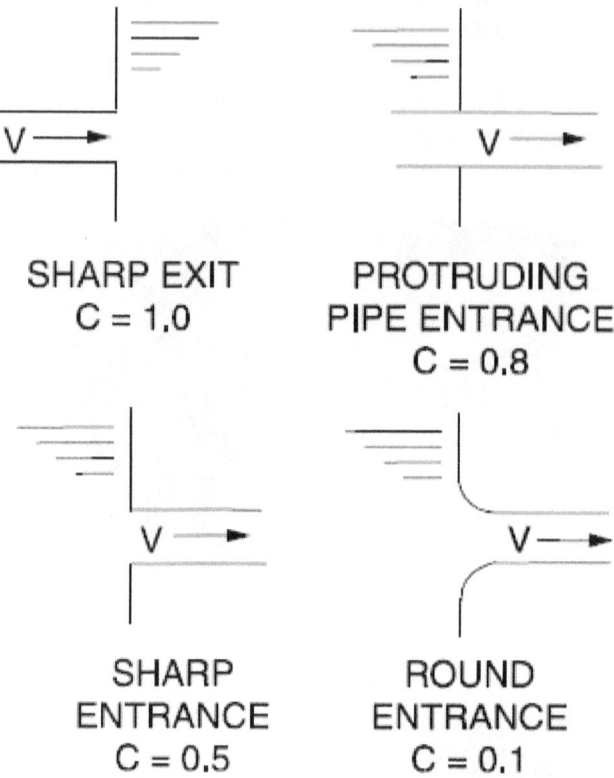

Figure 44. Head Loss Coefficient for Entrance or Exit

Top left: you're going to lose one velocity head (C=1.0) when the air leaves the conduit. You might recover some, but not all, of this with a flare. Top right, bottom left, and bottom right: when you draw air in, you're going to lose 10% to 50% to 80% of one velocity head due to the entrance alone. This doesn't even include the fact that sucking the air from the great outdoors up to your inlet through a perfect, infinite, frictionless cone will cost you another velocity head according to Bernoulli's Equation. Engineers that wouldn't neglect these details when sizing a pump, toss them aside when sizing a fan. The physics is the same.

If the area or the density changes, the velocity head will change also. Sample calculations are on tab, Sheet4, of examples\fan.xls. Typical values for user inputs are at the top (bold blue text) and calculations are at the bottom (orange text). The density is fairly standard for air at atmospheric conditions and for the purposes here, assumed to be constant. The entrance and exit losses are similar to those in the previous figure. The fill loss is roughly proportional to the depth. The areas are reasonable. Each of the velocity heads is calculated, multiplied by the number lost, and then summed. The Excel Solver function is used to adjust the airflow (1,478,006 actual cubic feet per minute) so that the static pressure rise developed by the fan matches the losses (0.9930 inches of water, a common unit of measure for air flows and natural hold-over from the days of glass manometers). The error in pressure (required vs. supplied) is in bold red text and the Solver target value. The required motor power is calculated from the curves, which have been implemented as VBA macros in the spreadsheet for convenience).

	A	B
1	Fan Sizing	Example (constant density)
2		inputs
3	0.075	density [lbm/ft³]
4	75	inlet diameter [ft]
5	60	fill diameter [ft]
6	30	fan diameter [ft]
7	6	fill depth [ft]
8	0.50	entrance loss [heads]
9	0.25	exit loss [heads]
10	2.75	drift eliminator loss [heads]
11	16	pitch angle [°]
12		calculations
13	35.0	fill loss [heads]
14	4417.9	inlet area [ft²]
15	2827.4	fill area [ft²]
16	706.9	fan area [ft²]
17	1,478,006	flow [ACFM]
18	0.9930	dPfan [in.H2O]
19	5.6	inlet velocity [ft/sec]
20	8.7	fill velocity [ft/sec]
21	34.8	fan velocity [ft/sec]
22	0.0070	inlet velocity head [in.H2O]
23	0.0170	fill velocity head [in.H2O]
24	0.2722	fan velocity head [in.H2O]
25	40.50	sum of losses [heads]
26	0.9930	sum of losses [in.H2O]
27	0.0000	pressure error
28	304.7	motor [bhp]

Figure 45. Sample Fan Calculations

Once a device containing a fan, such as a cooling tower, is built, the fans are often adjusted for optimum performance, which might include cooling and also electrical power consumed by the motors. This is done by *pitching* the fans just enough to meet the cooling requirements, as greater pitch almost always means more power consumed. This same spreadsheet has a button and corresponding VBA macro to create a table of flow and power vs. pitch angle. The results are shown in this next figure.

41

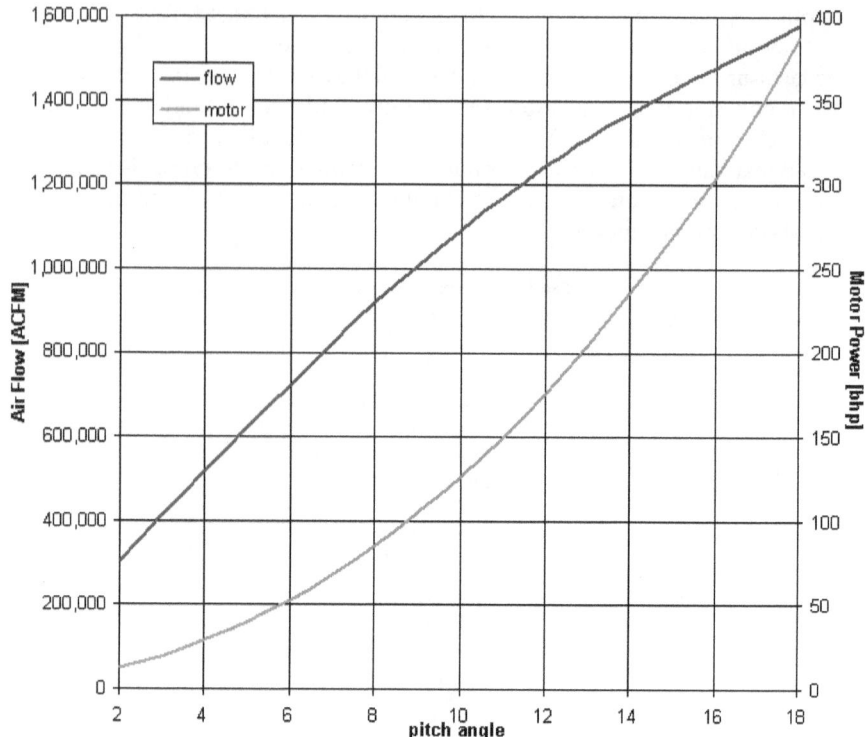

Figure 46. Results of "Pitching" the Fan

We can also set motor horsepower and vary both airflow and pitch angle to match the pressure drop and so create a somewhat different pair of curves. This requires two discrepancies (delta-P and delta-bhp) to be zero. The Solver easily accomplishes this by taking the sum of the squares and seeking a minimum rather than seeking a single zero, as before. This is on the next tab (Sheet5) in the same spreadsheet (fan.xls). The VBA macro steps through the solutions and fills in the table, which is shown in the following graph.

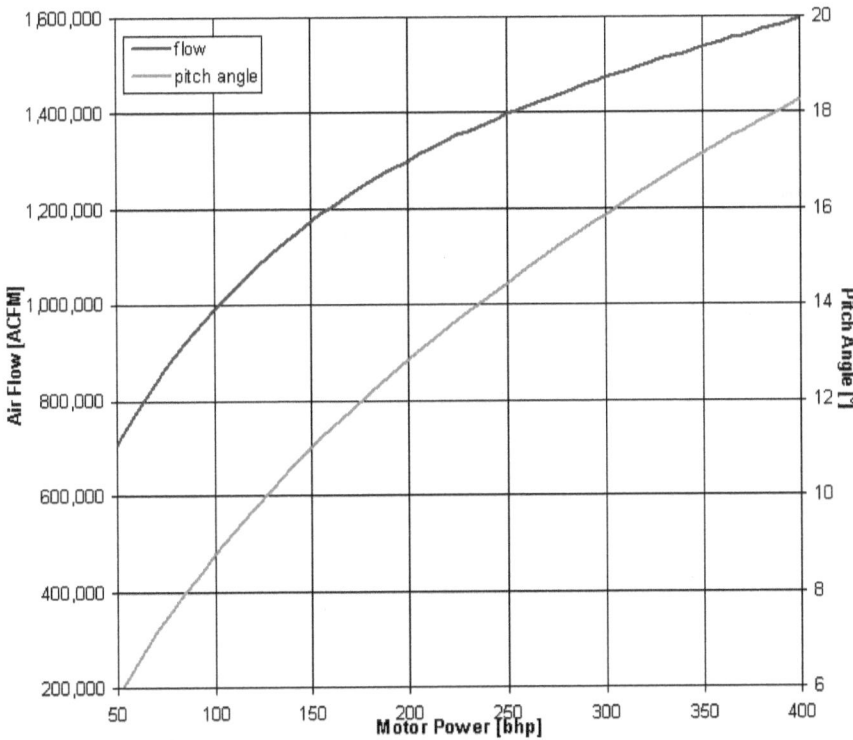

Figure 47. Motor Power as the Independent Variable

<u>Fan Motor Power</u>

It is worth mentioning here that some confusion exists over motor power when using SI units. There is no such confusion when using English units, as bhp (brake horsepower) clearly means mechanical power delivered by a rotating shaft and kW (kilowatts) clearly means electrical power consumed. The two are not the same. The electrical and mechanical efficiency of a motor and gearbox, if there is one, will be something less than 100%. The distinction is sometimes made between kWe (kilowatts *electric*, that is, one kilo-Joule/second) and kWm (kilowatts *mechanical*, that is, one kilo-Newton-meter/second). Do not make the costly mistake of presuming a motor that says 1000 kW on it somewhere will actually *deliver* 1341 bhp because this may simply be the nominal power *consumed*. Single-phase 110V shop motors of approximately ¼ to ¾ hp have an efficiency of about 45%. Larger (100+ bhp) three-phase motors may have an efficiency closer to 75%. Fans don't care about electrical power consumed, only mechanical power delivered by a rotating shaft.

43

Chapter 4. Turbochargers

The design of most turbochargers is somewhere between axial fans and centrifugal compressors and so we consider them in that order. Turbochargers rotate much faster and are much smaller than the axial fans we considered in the last chapter. While there are many nuances and seemingly minor shape details that impact the performance, these devices have much in common. A typical vintage automotive type is shown dismantled below. The impeller in the middle is the one that handles the air. The other handles the exhaust.

A variety of turbochargers are available online from several manufacturers. Some simply state the most common applications, while others provide more detailed information. We are interested in the details and so one is selected for which performance curves are readily available from the manufacturer, in this case Garrett™. The particular model is GTX4594R, a 70 mm dual ball bearing, single shaft unit, said to be capable of feeding compressed air to an engine producing up to 1250 hp (presumably net power at the flywheel, not on the pavement at the rear wheels).

The published performance curves for this device are shown on the next page. These can be readily found with a web search and are an excellent example of such curves, which may be difficult to get for many similar devices. These curves show contours of constant efficiency on a grid of pressure ratio vs. flow. Lines of constant rotational speed (rpm) are also shown.

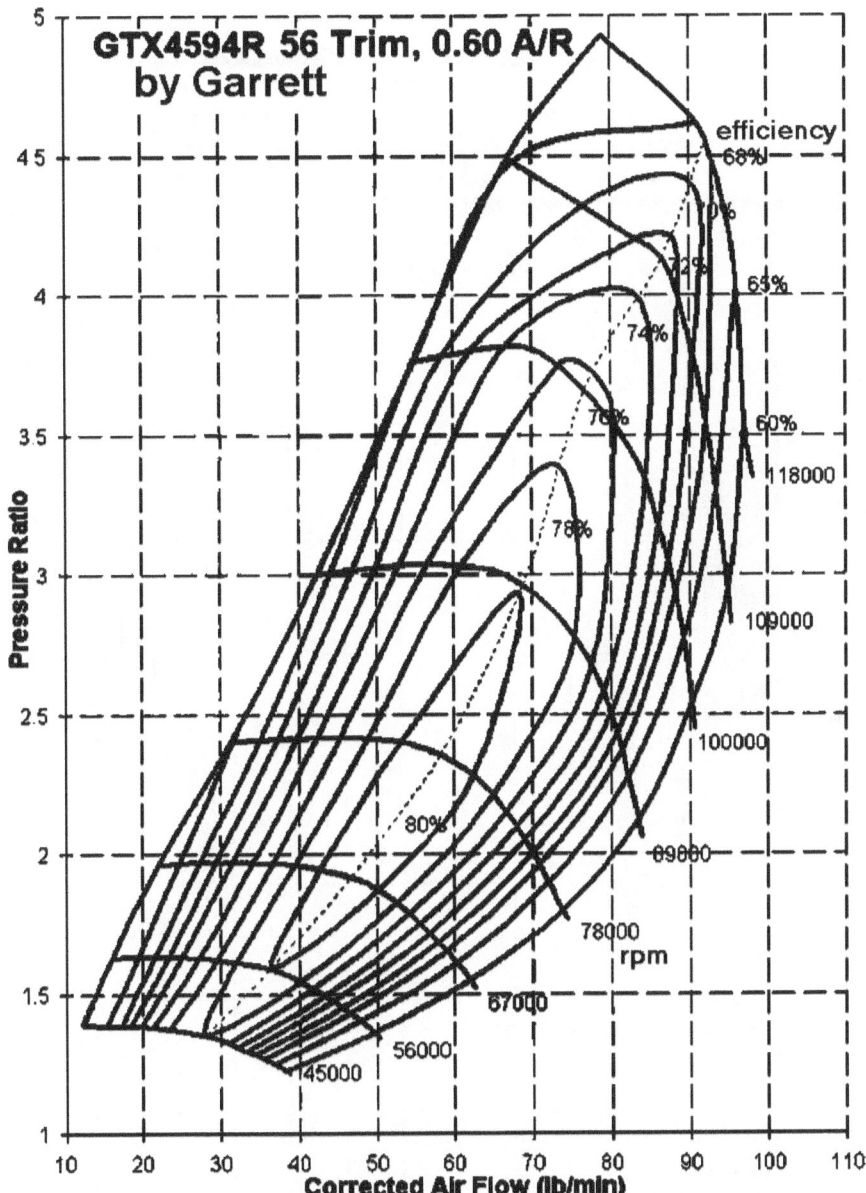

Figure 48. Garrett's GTX4594R Turbocharger Curves

Notice that the 80% and 78% efficiency contours are closed and the others might be also, if the information extended far enough. These contours are much like a topographic map with the dotted trail indicating the ridgeline. The pressure ratio is the outlet pressure divided by the inlet pressure. The corrected air flow is referenced to the design or baseline operating conditions. That this is a mass flow rate (lbm/min) and not a volumetric flow rate (CFM or ACFM) is also of primary concern to the designer, as internal combustion engines consume air by mass and not by volume. While volumetric flow rate is interesting, this isn't what combines with the fuel to release heat and ultimately power. Chemistry determines the air-to-fuel mass flow ratio in a combustion engine. You will find the curves (digitized using the free tool available at the address in the Forward) in spreadsheet examples\turbo.xls.

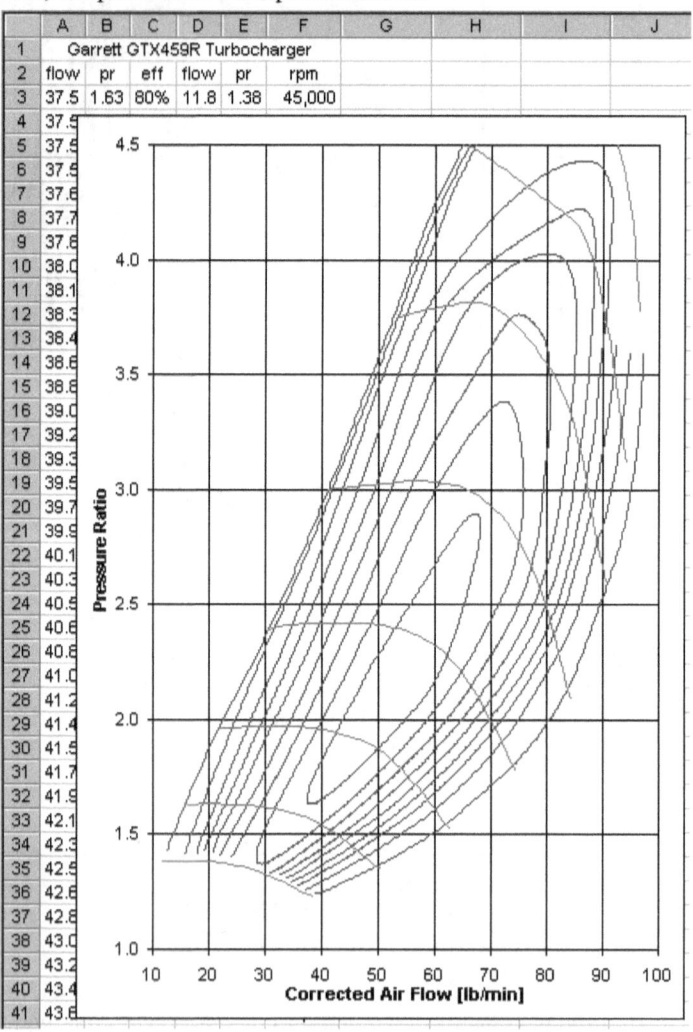

	A	B	C	D	E	F	G	H	I	J
1	Garrett GTX459R Turbocharger									
2	flow	pr	eff	flow	pr	rpm				
3	37.5	1.63	80%	11.8	1.38	45,000				
4	37.5									
5	37.5									
6	37.5									
7	37.6									
8	37.7									
9	37.8									
10	38.0									
11	38.1									
12	38.3									
13	38.4									
14	38.6									
15	38.8									
16	39.0									
17	39.2									
18	39.3									
19	39.5									
20	39.7									
21	39.9									
22	40.1									
23	40.3									
24	40.5									
25	40.6									
26	40.8									
27	41.0									
28	41.2									
29	41.4									
30	41.5									
31	41.7									
32	41.9									
33	42.1									
34	42.3									
35	42.5									
36	42.6									
37	42.8									
38	43.0									
39	43.2									
40	43.4									
41	43.6									

Figure 49. Digitized Turbocharger Curves

48

In order to use the curves for calculations, we need a map of efficiency and rotational speed. These have already been created (using TP2, which is available free at the location listed in the Forward). Two Tecolot™ data and layout files have also been created (turbo1.dat, turbo1.lay, turbo2.dat, and turbo2.lay), also in the examples folder of the online archive accompanying this text. The efficiency surface is shown in this next figure:

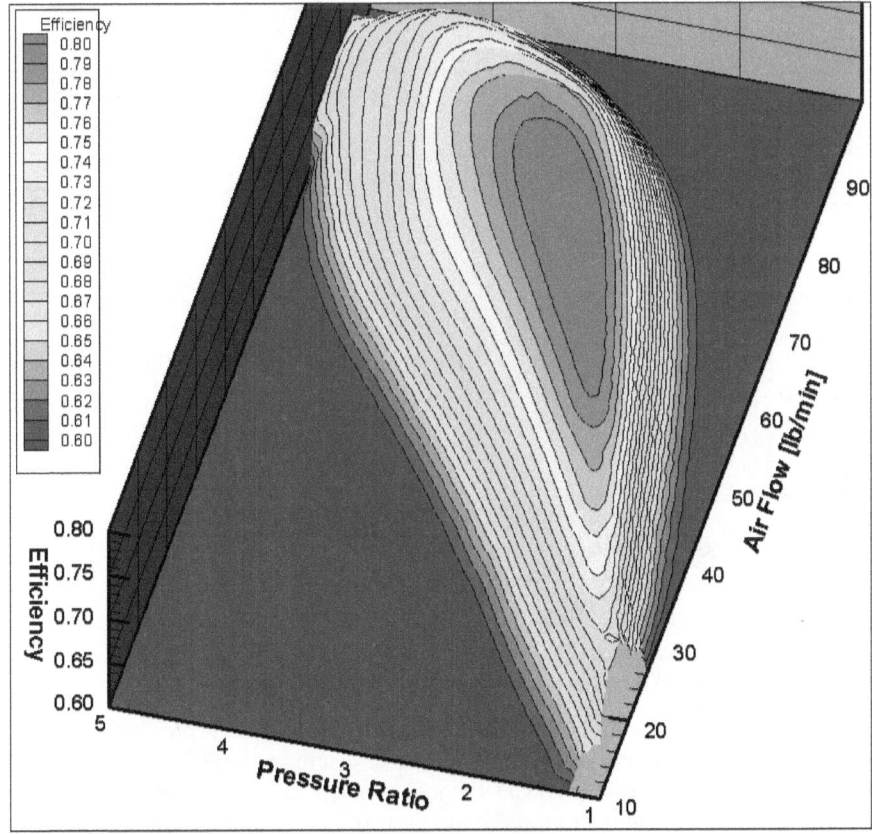

Figure 50. Turbocharger Efficiency Surface

Approximation of this surface using multivariate regression is not practical; therefore, we will use lookup tables created with inverse distance interpolation. We read the tables from the Tecplot™ data files so as to not waste space having the same information in two places. The program (turbo.c) we will use to make calculations is in the same folder. There is also a little batch file to compile it (_compile.bat). The code is ANSI standard C and should compile on any operating system.

The rotational speed map is shown in this next figure:

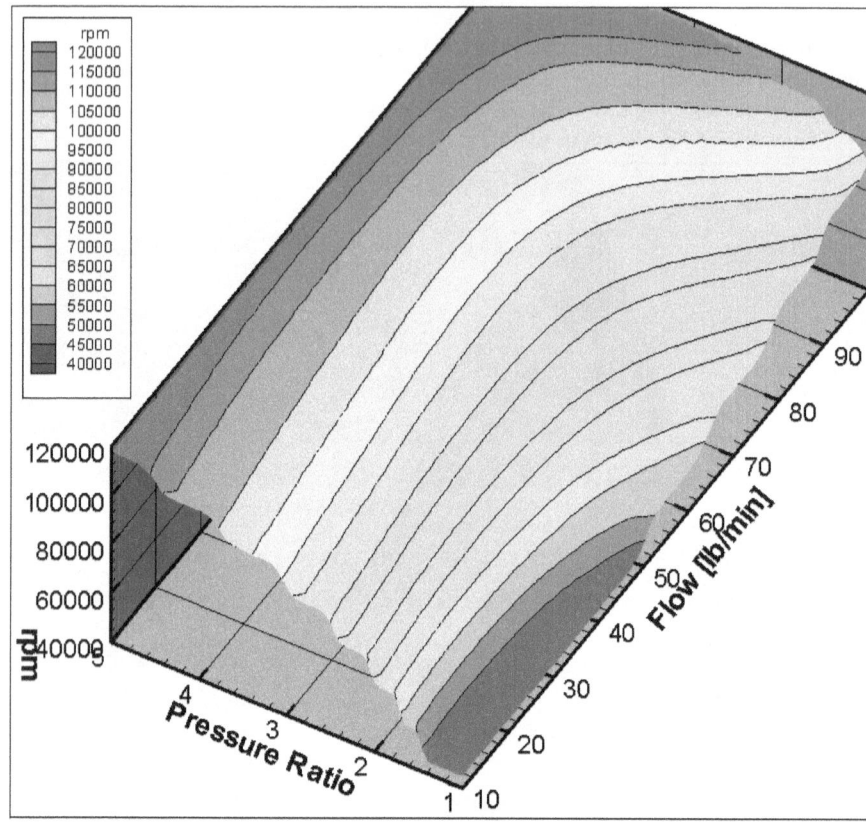

Figure 51. Turbocharger Rotational Speed Surface

Even after smoothing both of the surfaces (efficiency and rpm) have ripples. These unwanted artifacts arise from the coarse discretization (lumpiness) of the contours. Of course, real physical devices don't operate this way. Except for notable boundaries and operational limitations (e.g., vibration, excessive bearing load, cavitation, etc.), most machines smoothly transition from one state to the next (e.g., rotational speed, volumetric or mass flow rate, developed pressure, temperature, etc.).

We have six prospective operating curves for this turbocharger corresponding to six throttle positions. These are shown as four thick green and two thick brown curves on top of the previous graphic in the same spreadsheet.

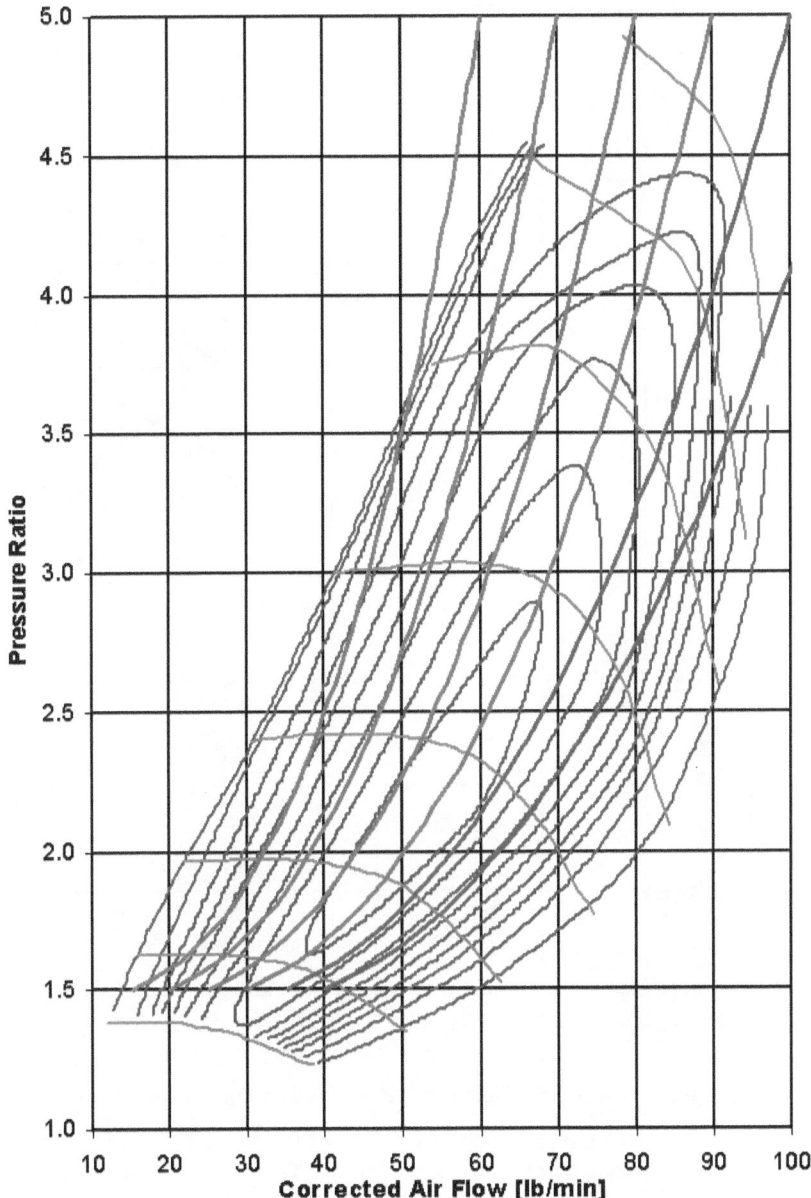

Figure 52. Prospective Operating Curves

These are pressure vs. flow curves, forced to match at the lower and upper operational limits and having the gross trend of the few available data points. We can use the lookup table approach (turbo.c) or a multivariate regression to deduce rotational speed corresponding to each arc.

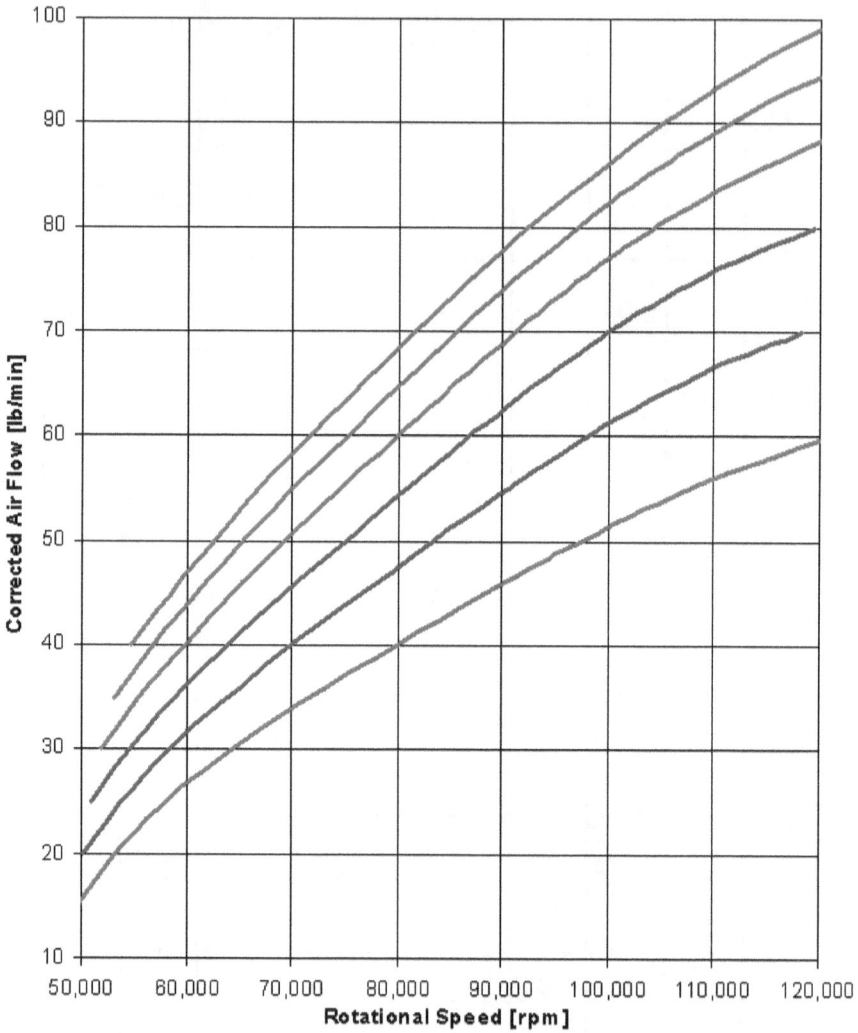

Figure 53. Rotational Speed for Prospective Operating Curves

Note that the two curves on the right side of the preceding graph have now appeared in the midst of the other four curves. This behavior arises from the mound-like 3D surface of the curves seen in Figure 50. The first four curves are on the left side of the hump and the last two are on the right side of the hump.

52

The pressure ratios corresponding to the same prospective operating curves are shown in this next figure:

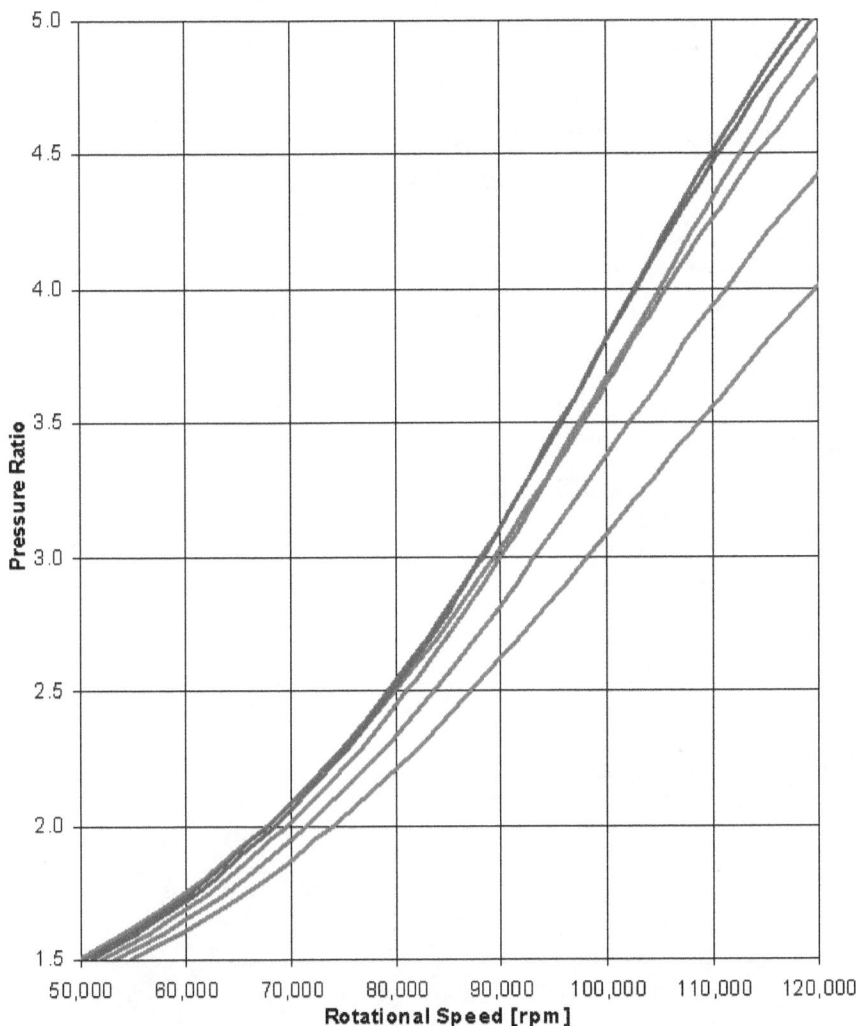

Figure 54. Pressure Ratios for Prospective Operating Curves

There is nothing unusual about this figure. The first four curves are below the last two so that the order is preserved.

The efficiencies corresponding to the same prospective operating curves are shown in this next figure:

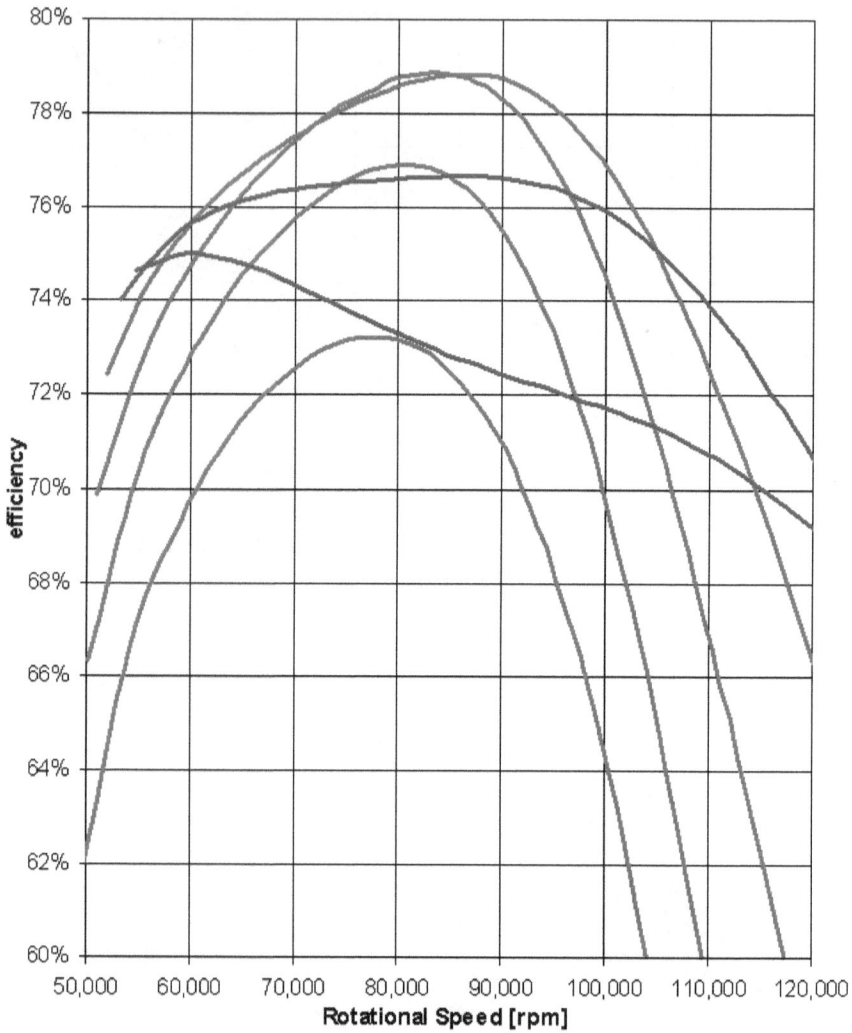

Figure 55. Efficiencies for Prospective Operating Curves

We see in this last figure that the last two prospective operating curves are definitely on a different part of the performance surface or map than the first four curves. This type of behavior is not at all unusual for a turbocharger. The interaction between the turbocharger and the other components of an internal combustion engine is far more complex than a naturally aspirated engine or one with a supercharger.

Chapter 5. Centrifugal Compressors

While a web search may turn up pictures similar to turbochargers, true centrifugal compressors are actually quite different in both design and operation. Turbocharger wheels look like a screw propeller, while centrifugal compressor rotors are shaped more like an old waterwheel or paddlewheel. The shaft bearings have very tight clearances, but turbochargers have relatively large clearances between the wheel and housing. Centrifugal compressor rotors have very small clearances. Turbocharger shafts are often quite small (perhaps 3/8" or 10 mm), while centrifugal compressor shafts are much larger (perhaps 3" or 75 mm). Turbochargers may be plumbed in series, but each unit has only a single stage. Centrifugal compressors often have multiple stages. The following figure is typical of these devices:

Figure 56. Typical Centrifugal Compressor

While a turbocharger might operate at a pressure ratio of 2 to 5, a centrifugal compressor with multiple stages may operate at a pressure ratio of 25 to 50. Not surprisingly, the performance curves are quite different also. We consider two similar compressors, both large industrial types, driven by variable speed electric motors. The flow is controlled by rpm and also guide vane position, which is reported in percent with 100% being "wide open" and 0% being completely closed. Typical data and curves may be found in the spreadsheet examples\centrifugal.xls.

	A	B	C	D	E	F	G	H
1	centrifugal compressor performance							
2	test points				performance curves			
3	GV	SCFM	psia	hp	GV	SCFM	psia	hp
4	100%	50,100	113	7130	100%	62,957	20	6260
5	100%	51,200	111	7210	100%	62,895	42	6311
6	100%	52,600	109	7280	100%	62,832	48	6362
7	100%	54,500	105	7350	100%	62,769	52	6413
8	100%	56,200	100	7390	100%	62,706	54	6460
9	100%	57,600	95	7390	100%	62,644	57	6503
10	100%	58,800	90	7350	100%	62,581	59	6543
11	100%	59,900	85	7280	100%	62,519	60	6581
12	100%	60,800	80	7170	100%	62,456	62	6616

Figure 57. Table of Centrifugal Compressor Performance

In this case we have test data and curves. The shape of the curves is known from experience with this type device. Data can be gathered from permanent or temporary instruments, the latter being more accurate and often specially calibrated. Measurements include pressure, temperature, torque, rpm, and flow. All of these are direct measurements, except flow. The only device capable of measuring flow directly is a Coriolis meter, which would not be practical for this application; therefore, flow is measured indirectly by the pressure drop across an orifice or nozzle. Torque is measured using a strain gauge, as illustrated below:

Figure 58. Strain Gauge Used to Measure Torque

Pressure operating curves for one particular device are shown in the following figure.

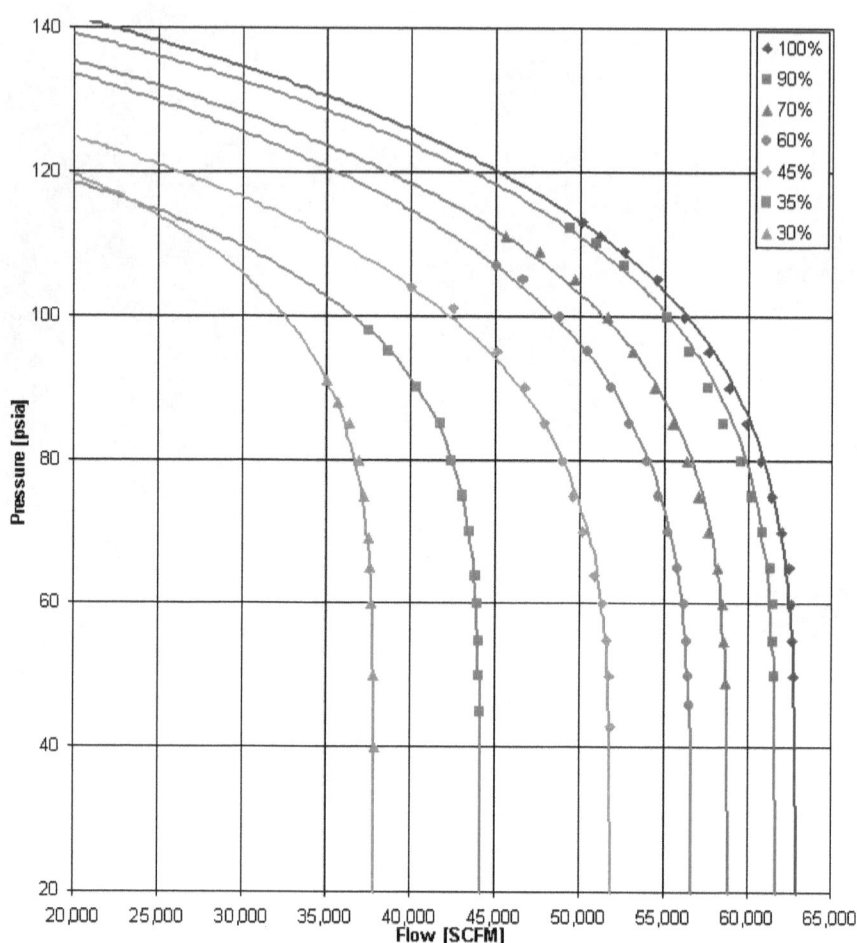

Figure 59. Operating Pressure Curves for Compressor 1

Not surprisingly, the pressure falls off rapidly with increasing flow, reaching some limiting value at each guide vane position, which is a characteristic of the design.

58

This compressor requires up to 7500 bhp (over 6000 kWe) to operate. It also requires continuous cooling, which is supplied by a recirculating water system and heat exchangers.

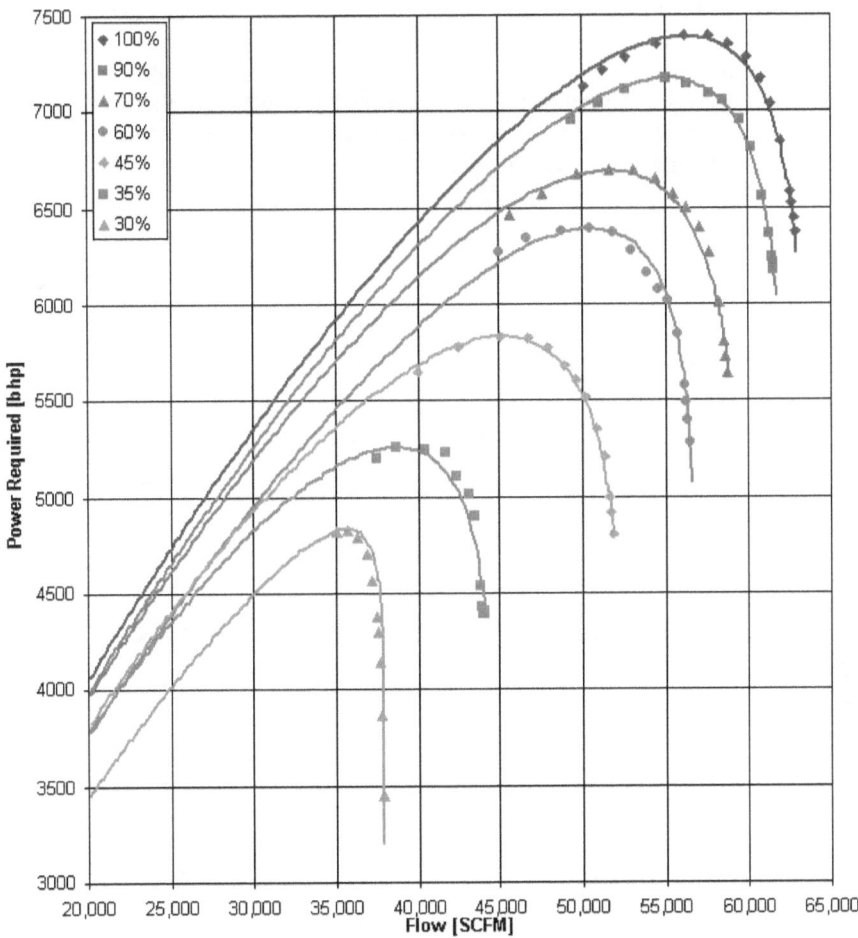

Figure 60. Operating Power Curves for Compressor 1

The required power rises sharply with increasing pressure and flow, as it is roughly proportional to this product (force per unit area times volumetric flow rate is equal to force time length per time, or power). These operating curves are essential for sizing the other equipment in this system and so the shape of these curves (see details in the spreadsheet) are of great interest.

The second centrifugal compressor we consider is smaller but operates at somewhat higher pressures. The test data for this second device was collected at two different times, several months apart. I requested the flow, pressure, and power. What I got the first time was the flow, pressure, and electric power consumed. Technicians had to install a torque sensor and measure shaft power, as this compressor is driven by a variable speed electric motor and power consumed does not directly indicate power delivered. For this reason, the pressure and power data are slightly out of sync at the top end where the curves fall off rapidly. First, the pressure curves:

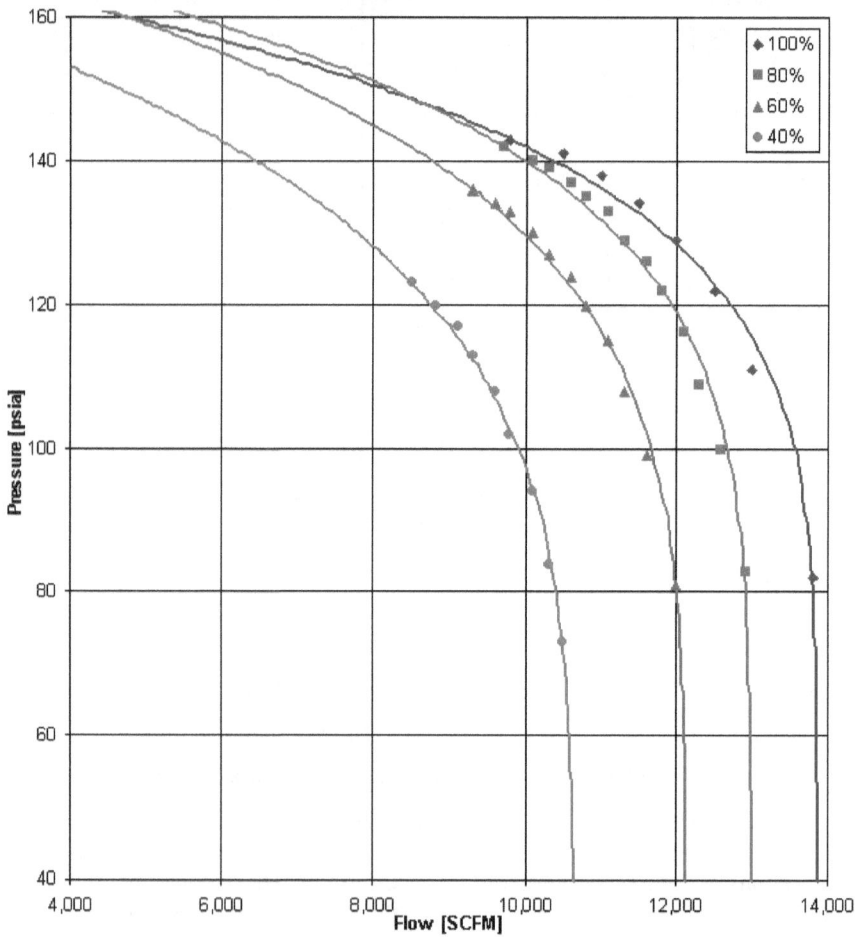

Figure 61. Operating Pressure Curves for Compressor 2

Discrepancies between the two data collections (i.e., testing events) is seen in the power curves:

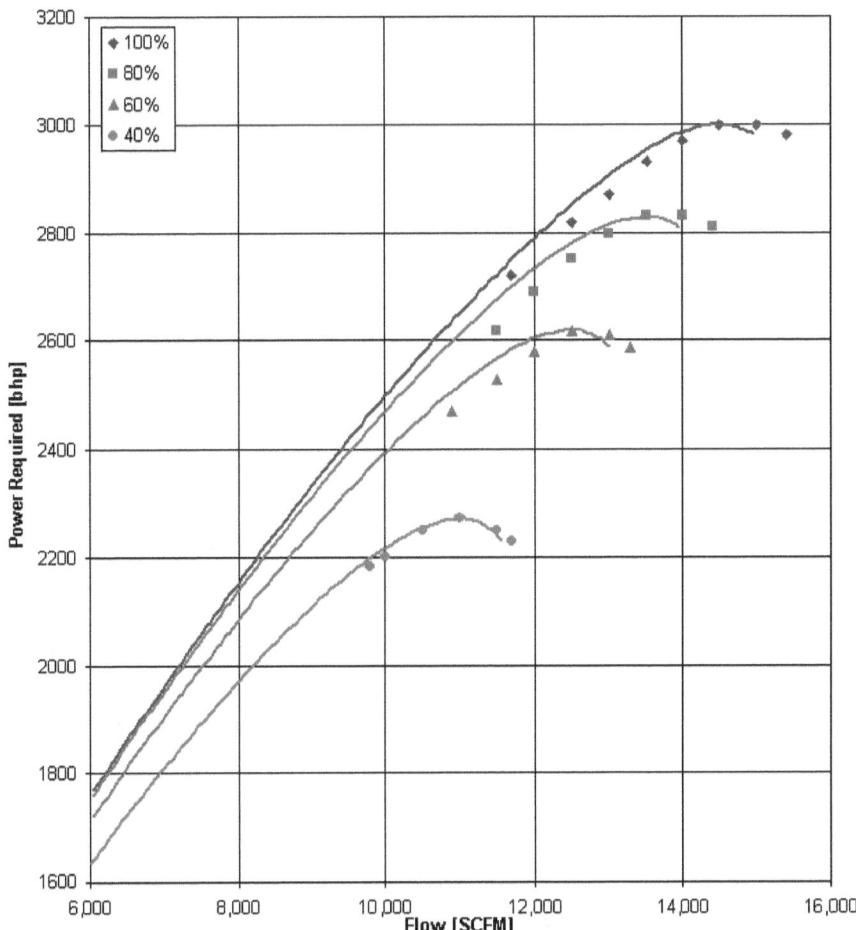

Figure 62. Operating Power Curves for Compressor 2

This experience reveals the importance of well planned testing, knowing what is required, and having the right instruments. It also illustrates the need to be clear when referring to mechanical vs. electrical power. It is usually much easier to measure electrical power. Don't rely too heavily on motor curves to derive one from the other.

Chapter 6. Screw Compressors

There are two basic types of screw compressors. The first of these is called after Rootes.[4] The rotors may have two or three lobes, as shown below:

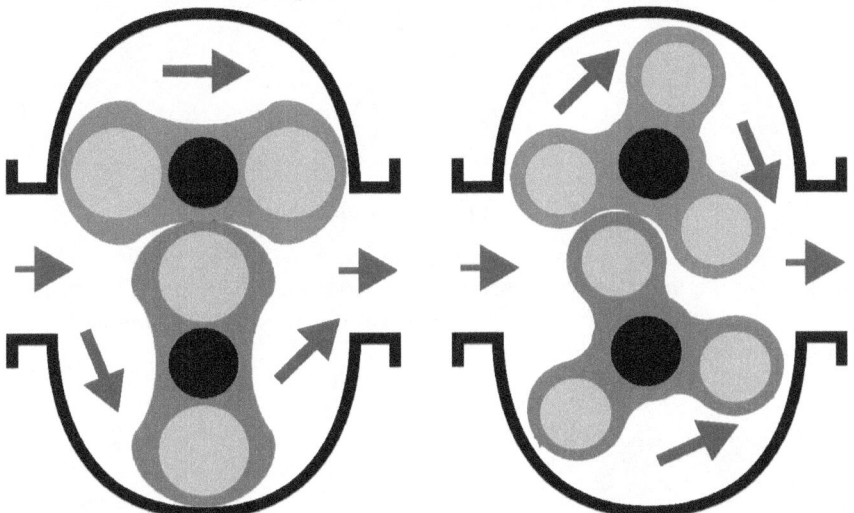

Figure 63. Two and Three Lobe Roots Compressor

The second type is called Lysholm.[5] Both come in symmetric (or mirror image) shape and also male/female varieties, as shown below:

---

[4] Rootes Motors Limited, a British company founded by William and Reginald Rootes.
[5] Alf James Rudolf Lysholm (1893–1973) Swedish engineer who developed the rotary screw compressor and the hydraulic torque converter.

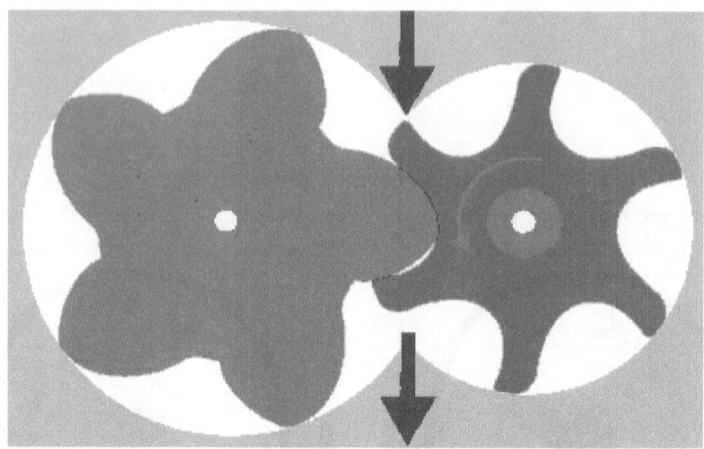

Figure 64. Male/Female Compressor Rotors

While the Rootes rotors are straight relative to the shaft, Lysholm rotors are twisted. Unlike the machines discussed in previous chapters, these are positive displacement devices. These are often used for superchargers and are a common site at drag strips and tractor pulls.

Figure 65. Typical Application for a Screw Compressor

Designs vary and several patents exist. While most have two or three lobes, some have four, five, or six. As these devices depend on very tight clearances, machining is quite expensive.

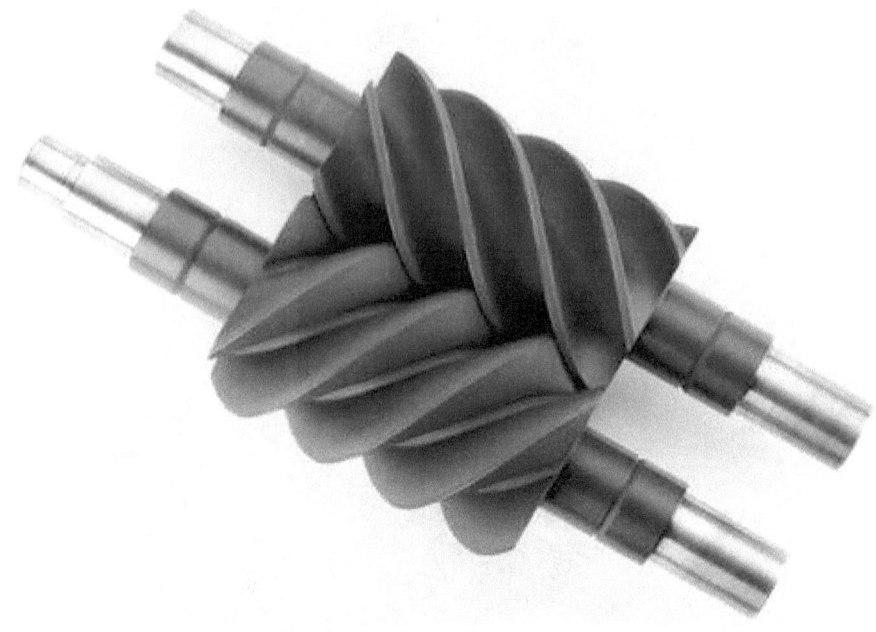

Figure 66. Twin Screw Rotors

Leakage and lubrication are a big concerns. Notice the Teflon™ seals in the rotors below.

Figure 67. Rotors with Teflon™ Seals

These fascinating devices have several uses in addition to supercharging internal combustion engines. Perhaps the most common application is to operate small industrial devices such as air drills and jackhammers. The operating pressure for such applications is about 160 psig (1200 kPa absolute). The efficiency is somewhat low, but the volume is high and a reciprocating compressor capable of delivering the same volume would be very inconvenient to drag about a construction site and maintain. Besides, efficiency is of little concern with such applications.

We first consider performance curves for a typical Rootes blower (see the first tab in supercharger.xls).

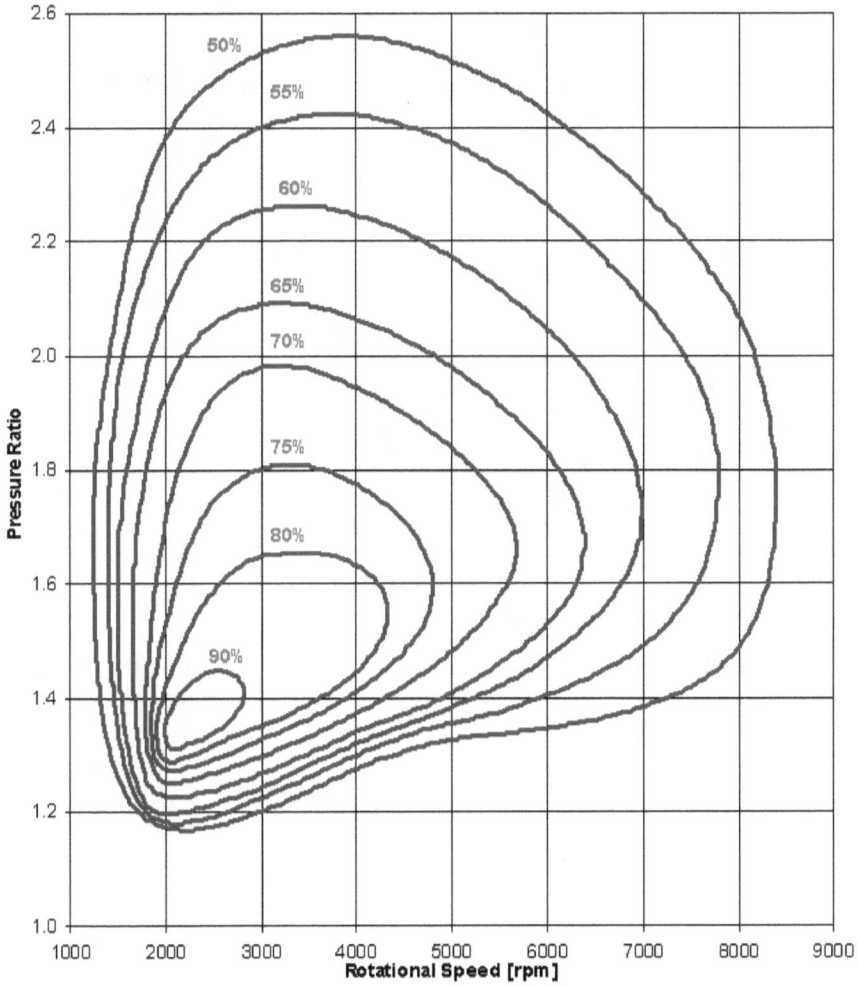

Figure 68. Rootes Blower Efficiency Curves

Notice that these contours are much wider and less distorted than those for the turbocharger. As mentioned previously, by comparison the operation of superchargers is much less complicated than turbochargers. The efficiency in 3D is shown in this next figure:

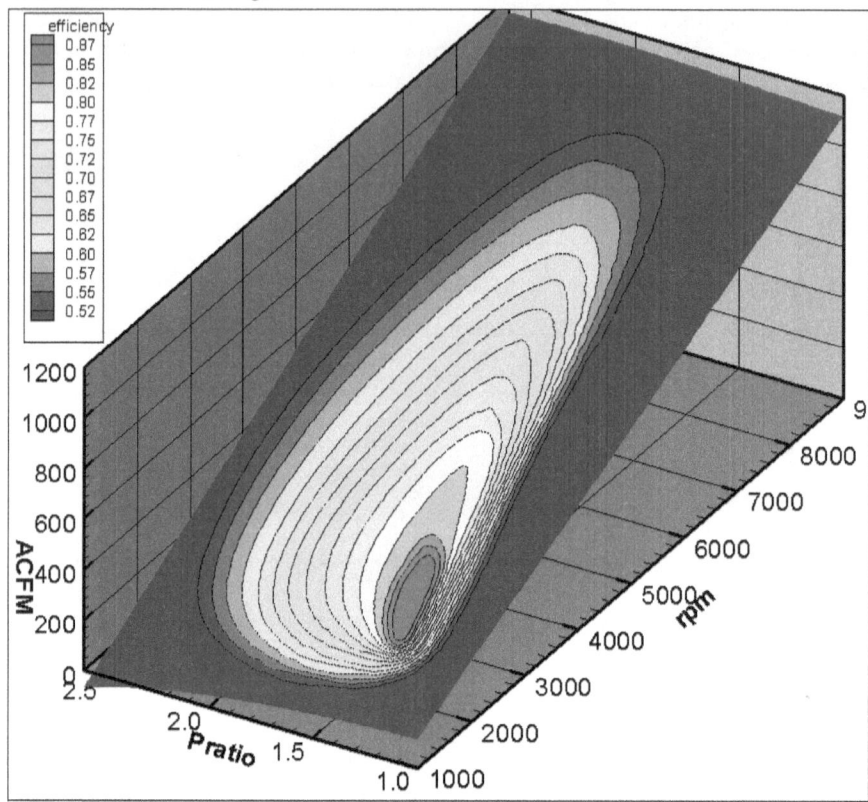

Figure 69. Rootes Efficiency Contours in 3D

The data for this device can be found in the online archive in the examples folder in supercharger.xls (on the first tab and also VBA macros) plus two files for use with Tecplot™: rootes.lay and rootes.dat.

The power required to drive this blower is shown in the figure below:

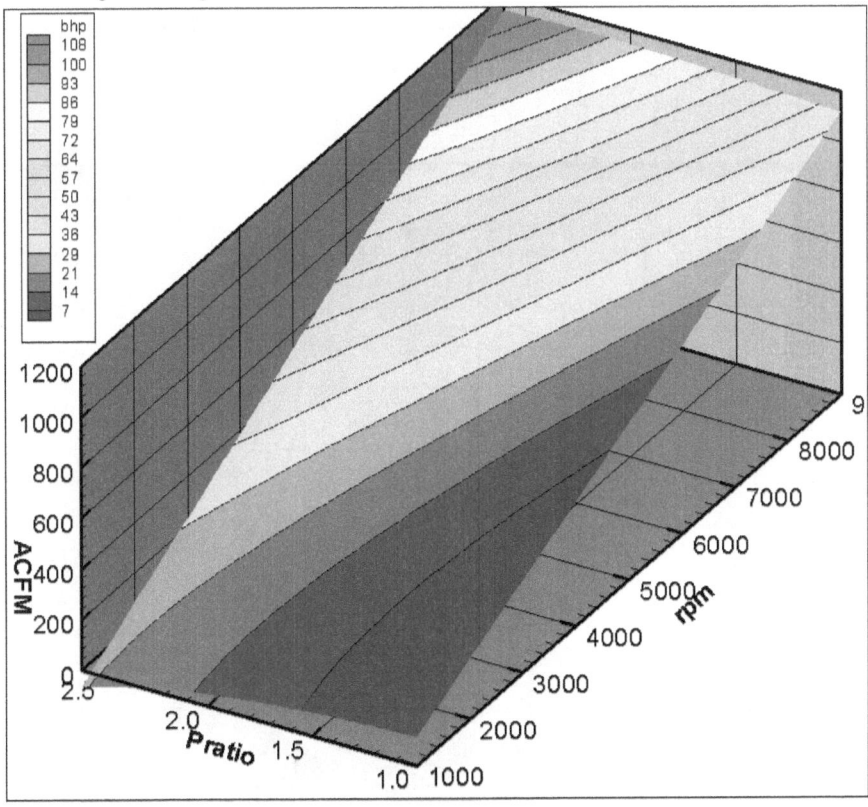

Figure 70. Rootes Power Contours in 3D

Lysholm blowers have somewhat different characteristics. They are significantly more expensive to machine. As the clearances are tighter and the shapes more accurate, they can often operate at higher rotational speeds. They don't necessarily generate higher pressures or larger volumes or have higher efficiencies, but any one and sometimes two or more of these advantages can be present in the same device. They may also be less likely to come apart, which is of particular concern in drag racing where the driver is practically sitting on top if the engine.

69

This next figure is typical of Lysholm blowers. You will find these curves in the spreadsheet supercharger.xls.

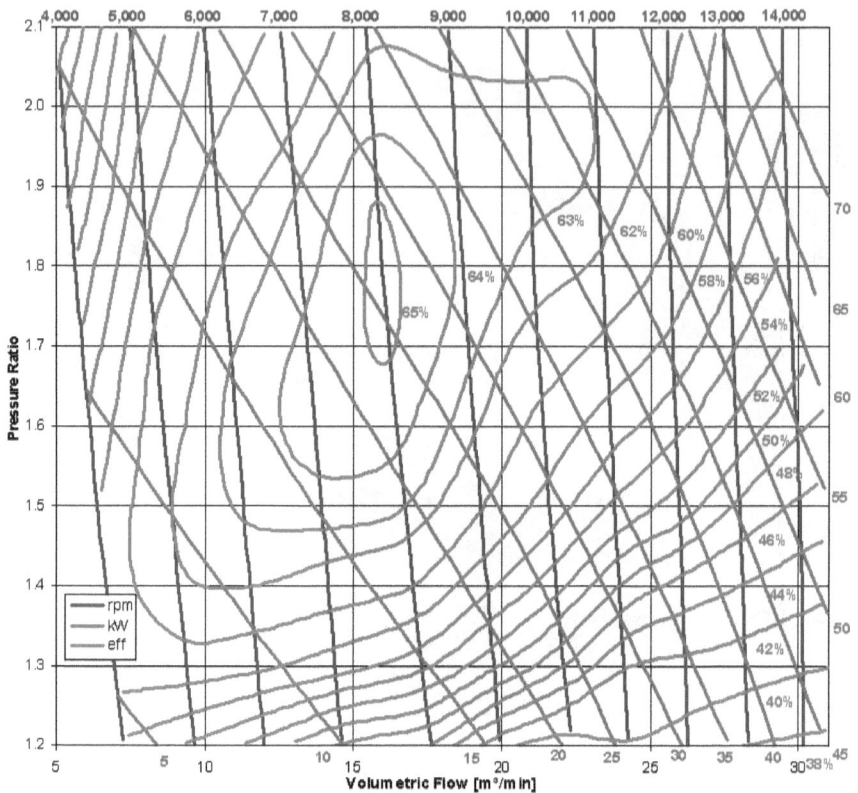

Figure 71. Lysholm Blower Curves

As for the Rootes, these curves are also much less distorted (squashed or stretched) than similar curves for a turbocharger. Notice that the power and volumetric flow curves in this and the previous figure for the Rootes blower are fairly straight, though diagonal in slope. Volumetric flow curves for the turbocharger were quite curved. This means that the performance of a supercharger is more proportional to the speed than the performance of a turbocharger, which rises quickly with increasing rpm. You can feel the turbocharging "kick in" at high rpm when driving a car so equipped. This is not ideal for drag racing, where you want maximum torque throughout the range of rotational speeds. You also want smooth, strong response when drag racing, not simply high end response, as one might want on the straightaway when road racing.

70

Chapter 7. Impeller Pumps

The first type of impeller pump we will consider is the vertical lift variety. These are used throughout large industrial plants, especially electric power generating stations, such as the ones I have spent the last forty years working around. While there are numerous smaller applications, the biggest ones are used to convey cooling water to and from a steam surface condenser and perhaps cooling towers. Typical flow rates range from 15,000 to 750,000 gpm (1 to 50 m³/sec). The head may range from 50 to 250 feet (15 to 75 m). As these pumps are designed to lift water, this translates into a rise in pressure of 20 to 100 psi (150 to 750 kPa). Not understanding the meaning of *head* and confusing the relationship between mass and force in SI units resulted in over-sizing by a factor of 10 the pumps at one power plant I know of. This mistake arose from neglecting to multiply by 9.8 m/sec².

Vertical Shaft Pumps

Such vertical lift pumps are often positioned at the plant intake or by the cooling towers, as illustrated in this first figure:

Figure 72. Typical Lift Pump Station

The man pictured here conveys the size of these devices. The electric motor and above ground portion may be 20 or 30 feet (6 or 9 m) tall. This is only part of the assembly. The pump housing and impeller are below ground, down in the

71

water, so as to assure net positive suction head (NPSH). If one of these pumps were to draw in a vortex of air or be run dry, permanent damage and injury may result. The motors may range from 1,000 to 35,000 bhp (745 to 26,000 kWm). I have spent days sitting on top of such a motor while it was running, taking measurements and collecting data. [shush… don't tell OSHA!]

This next figure shows the entire vertical shaft pump. This is a smaller unit, as you rarely see the big ones fully assembled.

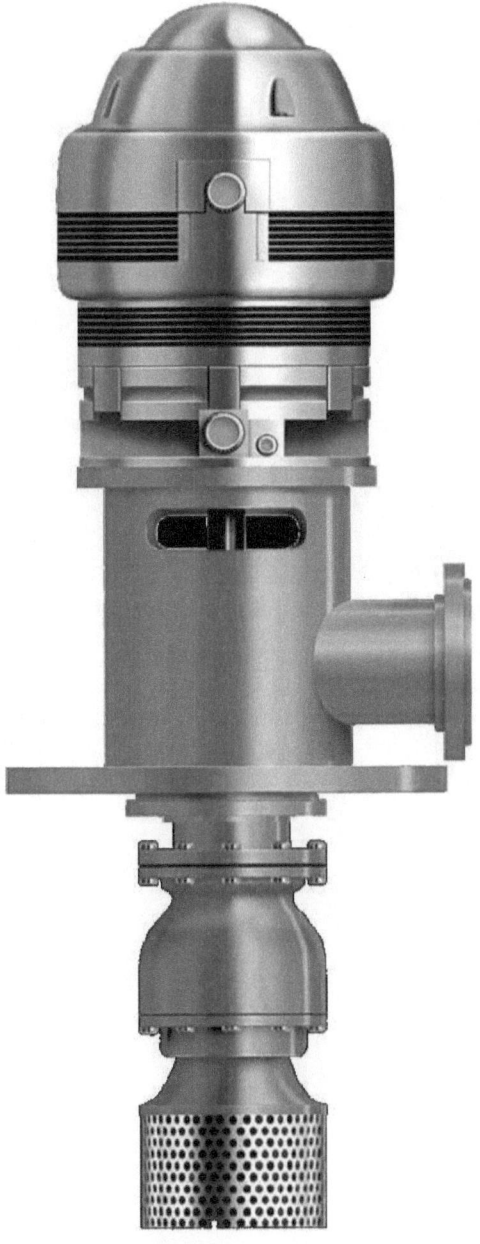

Figure 73. Vertical Shaft Pump Assembly

There may be one, two, or three of these per unit plus there may be more than one unit in the same plant so that a long row of these large pumps can be seen in some installations.

Figure 74. Row of Large Vertical Shaft Pumps

Steel grate on the floor near these pumps is a common site in plants. A ladder is shown and gives scale. The green pipes carry oil and cooling water to the bearings and motor. The red drums may contain more oil, cleaning solutions, or biocides, as slime and bacteria often grow in cooling water systems, especially if there is a cooling tower to aerate the water.

Vertical shaft pump impellers are usually simple in design and may look very much like the water pump impeller in a typical automotive engine—only much larger. The impeller shown in this next figure is 4 feet (1.2 m) in diameter.

Figure 75. Vertical Shaft Pump Impeller

You will find pump curves, curve fits, graphics, and calculations for a typical vertical shaft condenser cooling water pump in the online archive in folder examples and spreadsheet pump1.xls.

The head, efficiency, and power are shown in the composite figure below:

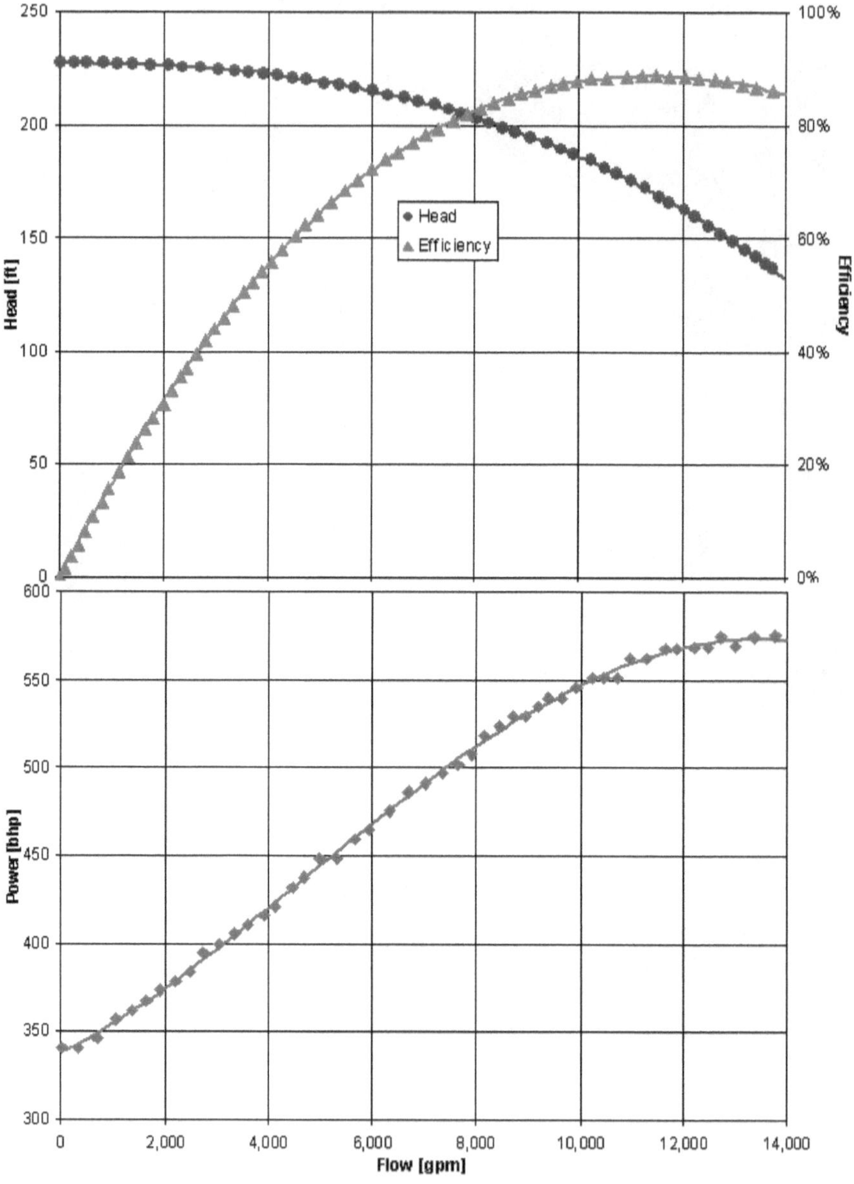

Figure 76. Vertical Shaft Pump Curves

At this plant there are three pumps in parallel. Multiple pumps provide backup and also versatility, as the plant does not always operate at full capacity.

The system resistance is a combination of pressure drops due to pipes, fittings, and the condenser water box. Included in this is the net vertical elevation change of the water. These combine to form a curve of head that increases roughly proportional to the square of the velocity. As the conveyance areas remain constant, the head is roughly proportional to the volumetric flow squared. The head provided for a single pump is shown in the previous figure. The head vs. flow for one, two, and three pumps is shown in the following figure along with the loss curve so that the intersections where the curves cross are where the system will operate with one, two, or three pumps.

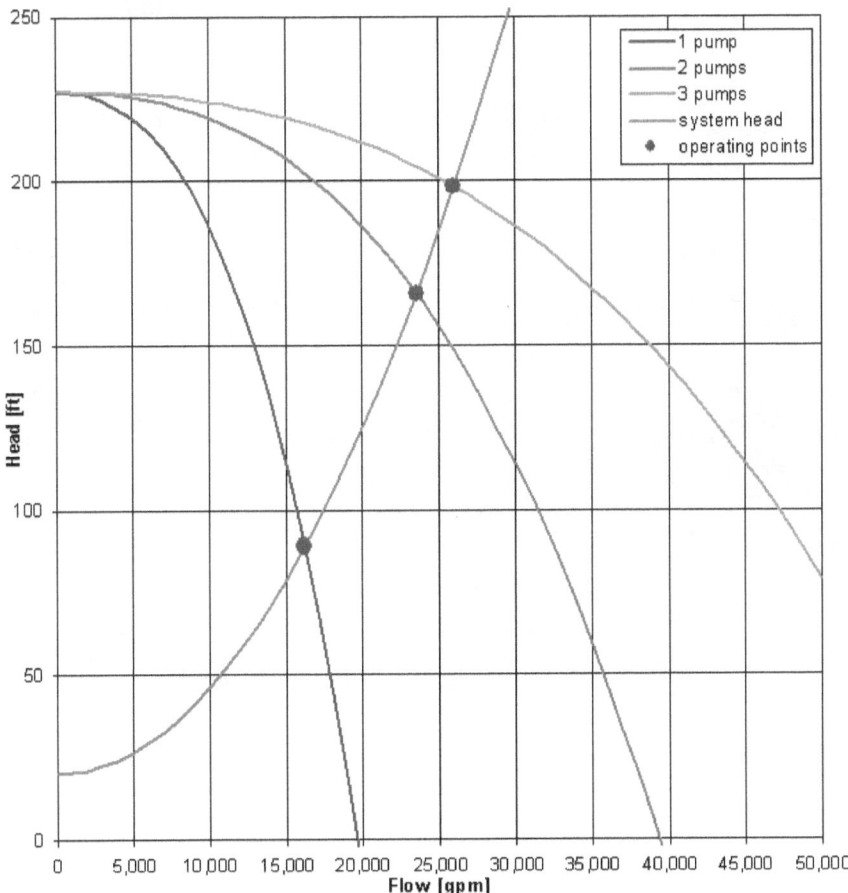

Figure 77. Operating Points for Three Vertical Shaft Pumps

A simple VBA macro (contained in the spreadsheet) calculates the flow using a bisection search, which is an efficient and reliable technique for solving the nonlinear equation (see Appendix D for details). Functions for the pump head and system loss are simple curve fits.

Horizontal Shaft Pumps

Horizontal shaft pumps typically seen in the same applications as these vertical shaft pumps may have somewhat higher efficiency or head or both and have impellers of a completely different shape—more like that of a Pelton wheel, as shown below:

Figure 78. Horizontal Shaft Pump Impeller

These impellers rotate inside a housing. These can be quite large and are usually below ground so as to assure net positive suction head on the pump, as we would not want them to cavitate. A typical industrial setting is shown in this next figure before the impellers and electric motors have been set into place.

Figure 79. Horizontal Shaft Pump Installation

As seen here, a large concrete pit is required for these pumps—something that was not necessary for the vertical shaft pumps, which have the motors high and away from the water. The pit must be kept dry too. So you see how there can be tradeoffs to consider, including: purchase, construction, operation, and maintenance.

In this case, the curves are quite similar, but that is not always the case.

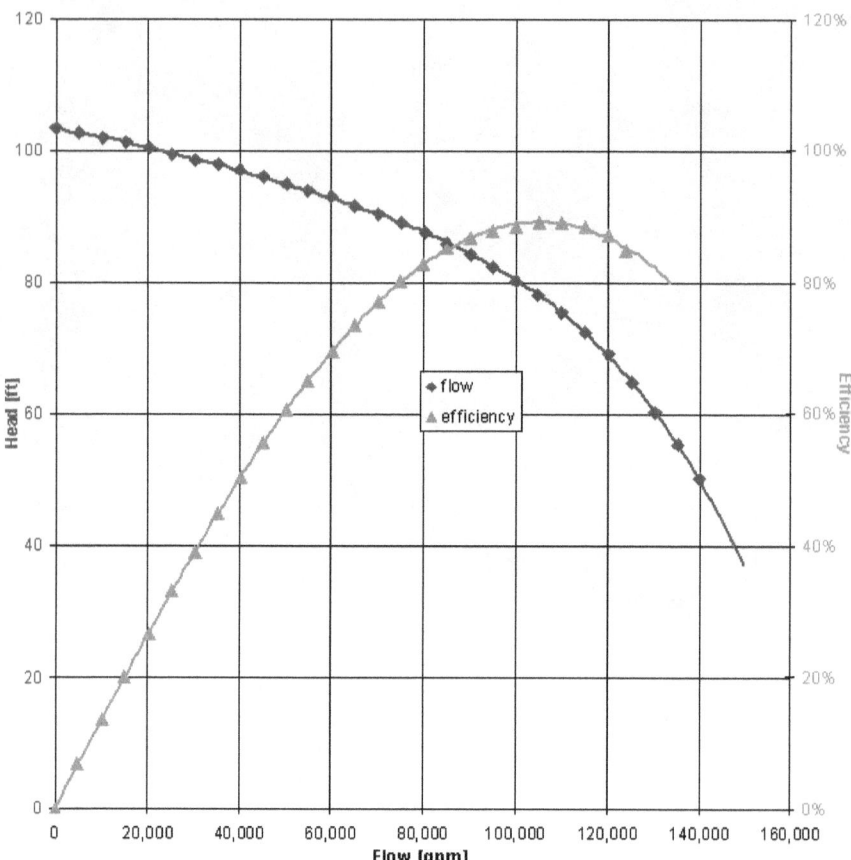

Figure 80. Horizontal Shaft Pump Curves

Not only is the NPSH critical for these particular pumps, but the electric motors must also be more flexible, as shown in these next curves.

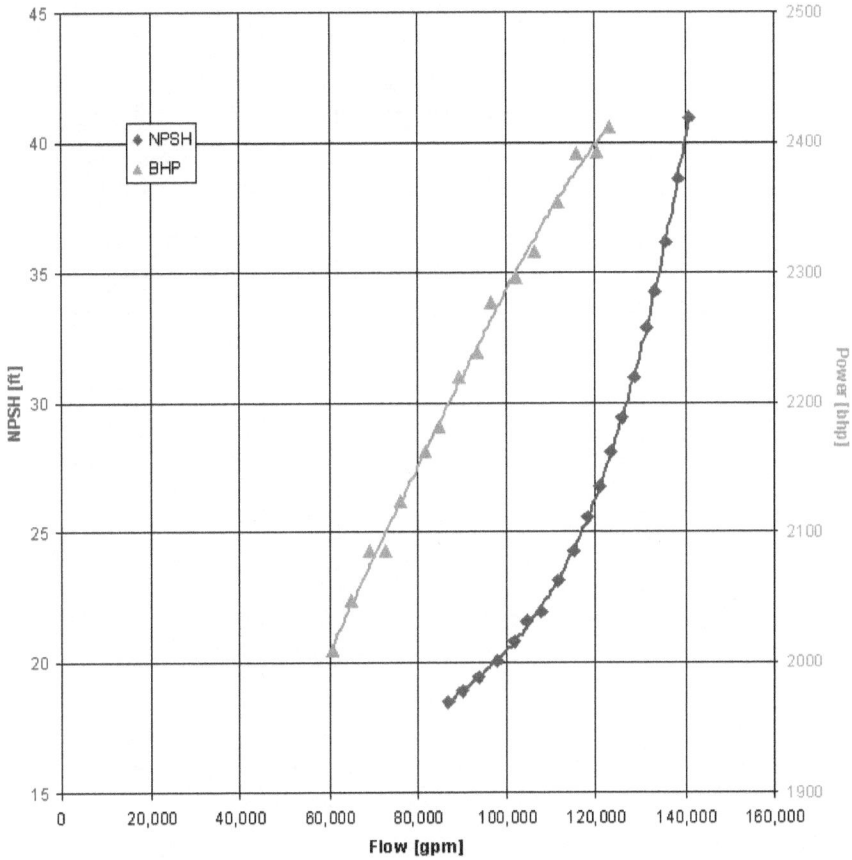

Figure 81. More Horizontal Shaft Pump Curves

Notice that the head and efficiency curves cover a significantly larger range of operation than the NPSH and power curves in the second set. While it may not be obvious, the range of operation for these pumps is more narrow than the vertical shaft ones shown previously.

We can readily find the operating point for one, two, and three pumps as before.

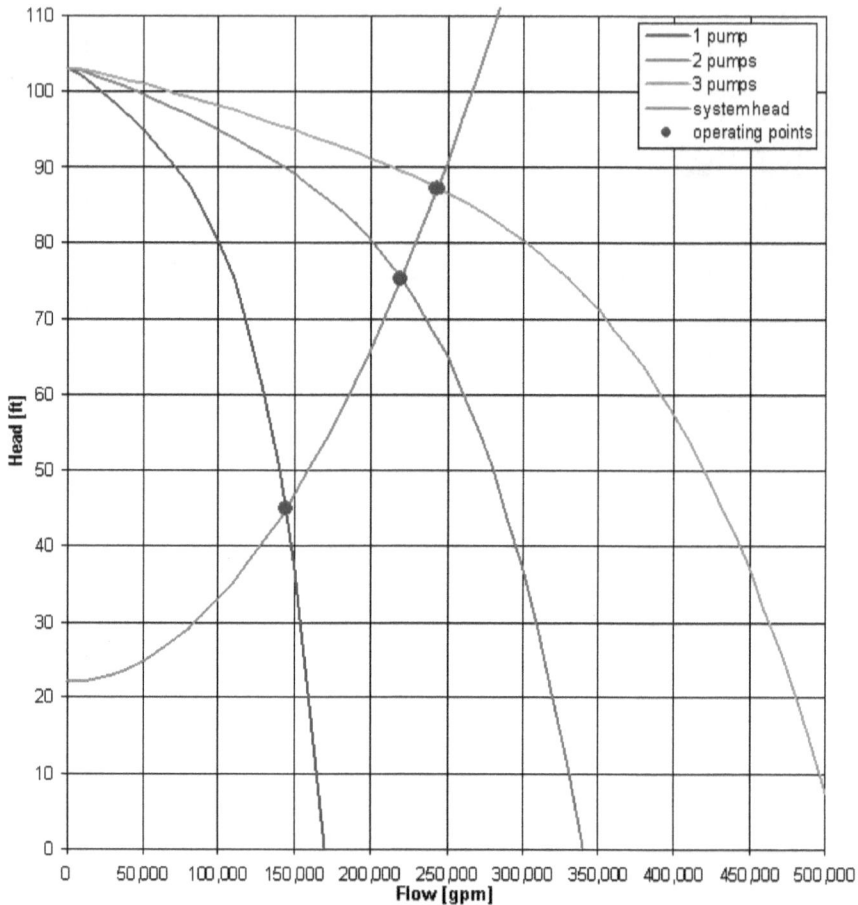

Figure 82. Operating Points for Three Horizontal Shaft Pumps

Another style of this same type horizontal shaft pump:

Figure 83. Horizontal Shaft Pump

Chapter 8. Boiler Feed Pumps

Boiler feed pumps (BFPs) operate at much higher pressures than vertical or horizontal impeller pumps. Just as centrifugal pumps operate at much higher pressures than turbochargers and have rotors with a significantly different shape, so do BFPs. In fact, the shape of BFP impellers is quite similar to that of some centrifugal pumps, as illustrated in Figure 56. A cut-away view is shown below:

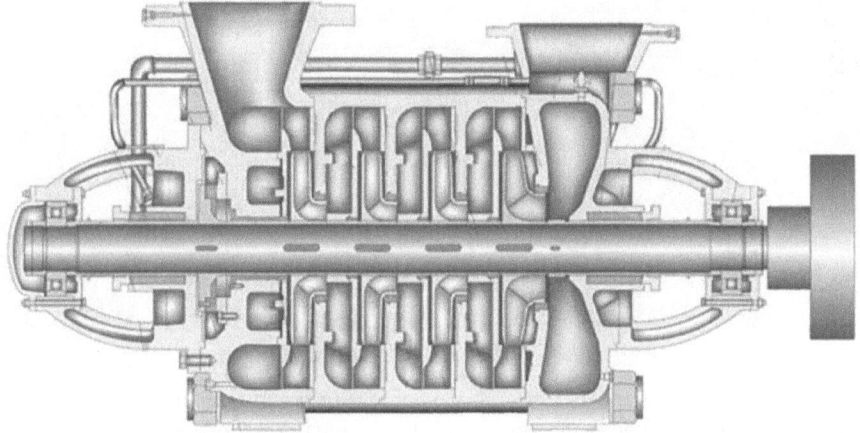

Figure 84. Boiler Feed Pump Schematic

These devices have much tighter clearances and rotate at a much higher speed. They also have multiple stages back-to-back in order to achieve the large pressure ratios. This is another similarity to centrifugal air compressors. Like centrifugal air compressors, BFPs are driven by variable speed electric motors or steam turbines. In all large coal plants and nuclear plants, the boiler feed pumps are driven by a steam turbine, as this is such an easily controlled source of variable speed torque.

A cut-away view is shown below. Note the glove for scale:

Figure 85. Boiler Feed Pump Cut-Away

These particular rotors are almost identical to a centrifugal air compressor, even of similar diameter and thickness. The centrifugal stresses are quite large and are of primary concern for the mechanical design. Continuous lubrication and in some cases water cooling are critical to operation.

The assembled units can be quite large and may be surrounded by massive plates and flanges. Such devices are often designed to be disassembled for maintenance; thus, may be held together with massive bolts, as shown in this next figure.

Figure 86. Large Boiler Feed Pump Assembly

The mounting brackets for this pump are also very substantial and vibration is continuously monitored to check for potential operational problems and needed maintenance.

Variable speed electric motors are usually coupled to the pump and both positioned on a single skid for transport and to maintain accurate alignment.

Figure 87. Boiler Feed Pump on Skid with Electric Motor

Multiple oil lines can be seen attached to this pump, as well as sensing lines and instruments cables. The oil system may be integral or separate, depending on the design and application. These pumps often require a different lubricant than nearby equipment, thus the need for a separate system. These motors are air cooled, but some are cooled by circulating water.

While it has always struck me as strange, most boiler feed pump curves are presented having flow in gpm. This may be a natural result of cold testing results or it may be a peculiarity of the industry. Sometimes the curves are presented in mass flow rate (lb/min).

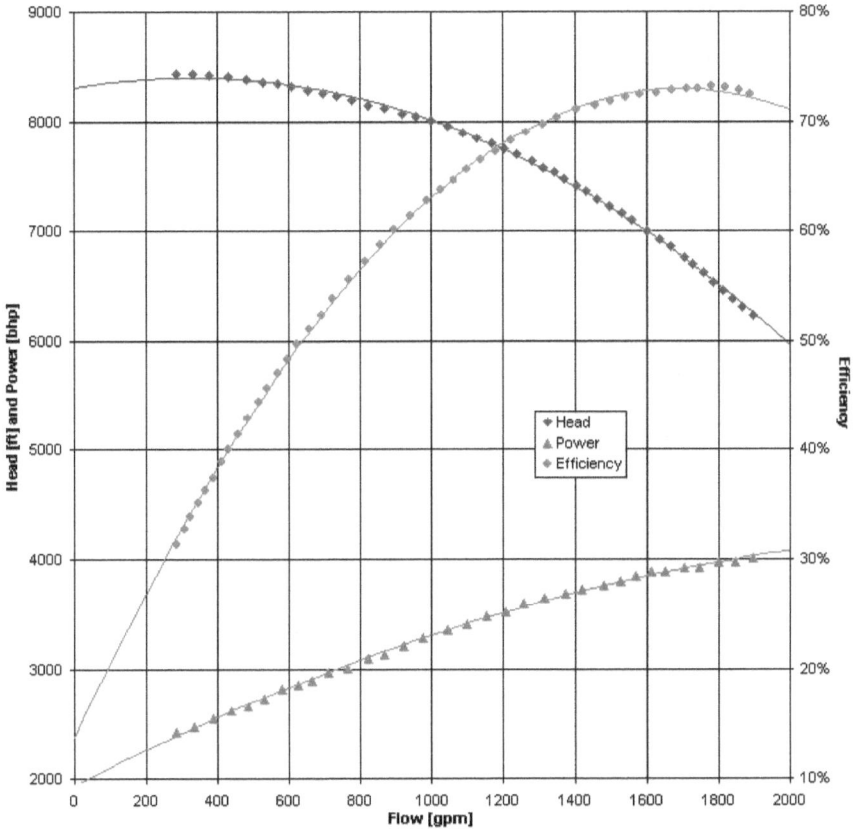

Figure 88. Boiler Feed Pump Curves

Boiler feed pumps can be quite large. Consider that two such pumps for a 900 MWe supercritical coal-fired power plant deliver 4,365,000 lb/hr at 4400 psi (550 kg/s at 30 MPa). Two boiler feed pumps at a typical 1200 MWe deliver 16,350,000 lb/hr at 1200 psi (2060 kg/s at 8.3 MPa).

Chapter 9. Squirrel Cage Fans

This is perhaps the most common type of fan, often used in heating, ventilating, and air conditioning (HVAC) applications. There are three designs, as shown in the figure below. We will only consider the last (forward-bladed), as the others are very similar to designs previously discussed.

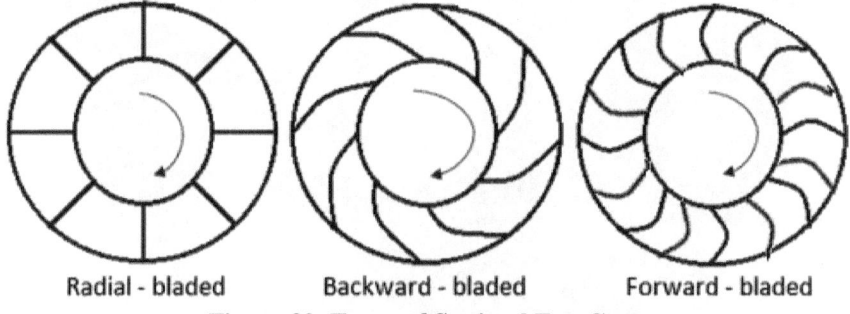

| Radial - bladed | Backward - bladed | Forward - bladed |

Figure 89. Types of Squirrel Fan Cages

Some small cages are made of plastic, while larger ones are metal. Blade detail can be clearly seen in this next figure.

Figure 90. Typical Forward-Bladed Cage

A typical assembled unit is shown in the figure below:

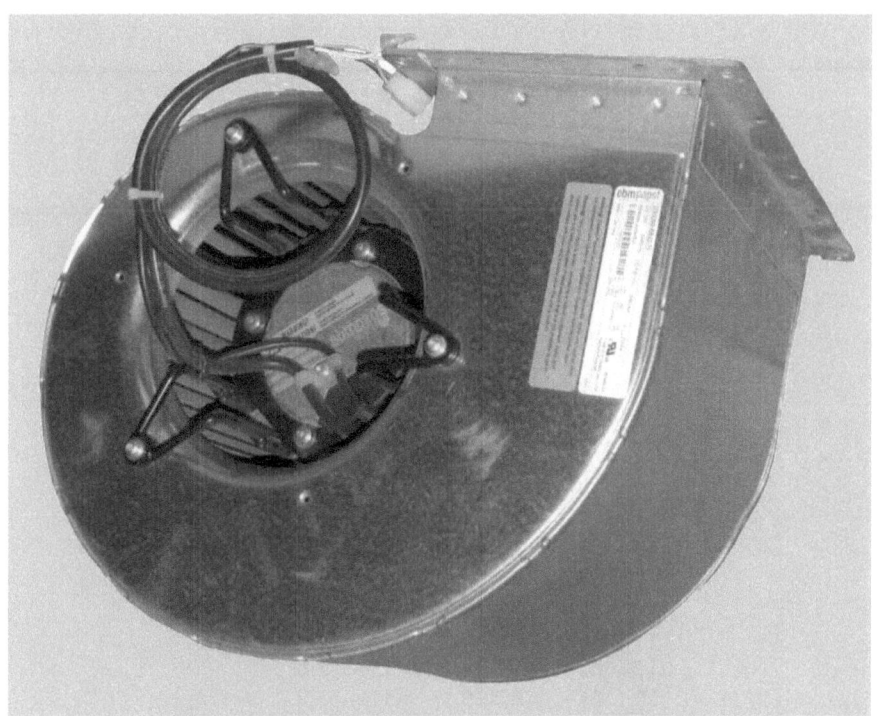

Figure 91. Typical Squirrel Cage Fan Assembly

The curves are a little different from the ones presented so far, although I do have at least one set of large condenser circulating water pump curves that have the same shape in spite of being vastly larger in size. The curves and data can be found in the online archive in folder examples and spreadsheet squirrel.xls. As before, we have several curves, including pressure developed vs. flow and system resistance vs. flow. Where these two meet is the operating point. This is illustrated in the spreadsheet with the intersection points identified in the chart.

The match column below indicates when the pressure developed by the fan is greater than, less than, or equal to the system resistance:

	A	B	C	D	E	F
1			squirrel cage fan			
2	cfm	in.H2O	eff	bhp	resist	match
3	0	2.12	0.0%	0.068	0.00	≫
4	5	2.06	7.8%	0.067	0.00	≫
5	10	2.00	14.9%	0.067	0.01	≫
6	15	1.95	21.4%	0.068	0.01	≫
7	20	1.90	27.3%	0.069	0.02	≫
8	25	1.86	32.6%	0.070	0.03	≫
9	30	1.83	37.4%	0.072	0.04	≫
10	35	1.80	41.7%	0.075	0.05	≫
11	40	1.78	45.6%	0.077	0.06	≫
12	45	1.76	49.1%	0.081	0.08	≫
13	50	1.76	52.2%	0.084	0.09	≫
14	55	1.76	54.9%	0.088	0.10	≫
15	60	1.77	57.3%	0.093	0.12	≫
16	65	1.78	59.4%	0.097	0.13	≫
17	70	1.81	61.2%	0.102	0.15	≫
18	75	1.83	62.7%	0.108	0.16	≫
19	80	1.86	64.0%	0.113	0.18	≫
20	85	1.89	65.1%	0.119	0.20	≫
21	90	1.93	66.0%	0.125	0.22	≫
22	95	1.96	66.7%	0.131	0.23	≫
23	100	1.99	67.2%	0.138	0.25	≫

Figure 92. Squirrel Cage Fan Spreadsheet

Pressure in inches (or centimeters) of water is quite convenient to measure with a length of clear plastic hose. I recall one performance test in a remote area. Several crates containing instruments were lost in shipping. The engineers had to make do with items purchased in the bazaar from a vendor selling automotive related parts and supplies, including lengths of plastic tubing for measuring pressure and a sack of coins for incremental weight. After the test, everything was shipped back to the laboratory and the procedures performed again, this time along side precision instruments. The experiment was surprisingly successful.

The curves and operating pointare show in this next figure:

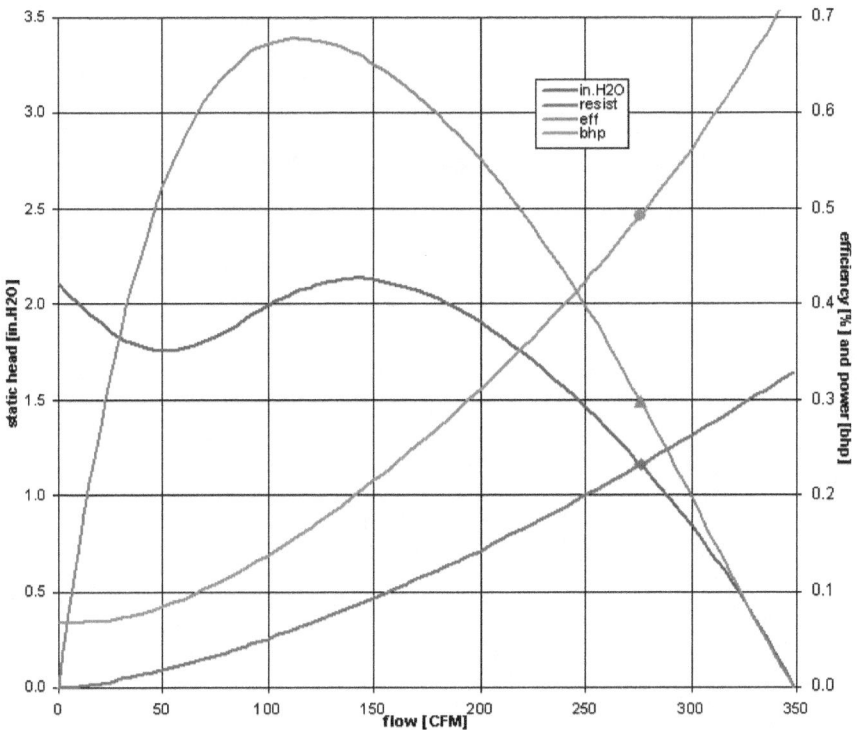

Chapter 10. Rubber Impeller Pumps

Outboard and inboard/outboard motors all use the same type of cooling water pump, only of different sizes. These consist of a rubber impeller turned inside an offset chamber that flexes the branches so the enclosed area is compressed and expanded, producing a flow and having positive suction, which is essential to self-priming. The impellers look like:

Figure 93. Water Pump Impeller

The slot accommodates a key in the rotating shaft. The entire device is remarkably simple, which increases reliability. Assured water flow is essential for cooling and loss for even a few seconds may mean catastrophic engine failure.

The chambers are also quite simple in design and vary from one model to another only in non-essential details. These may be cast aluminum or plastic.

Figure 94. Water Pump Housing

The housing is held in place by two, three, or four bolts. The drive shaft runs through the middle. Water enters from below through a plate and exits through the pipe slip connection on the top left of this figure.

Compression and expansion of the impeller can be seen in this figure:

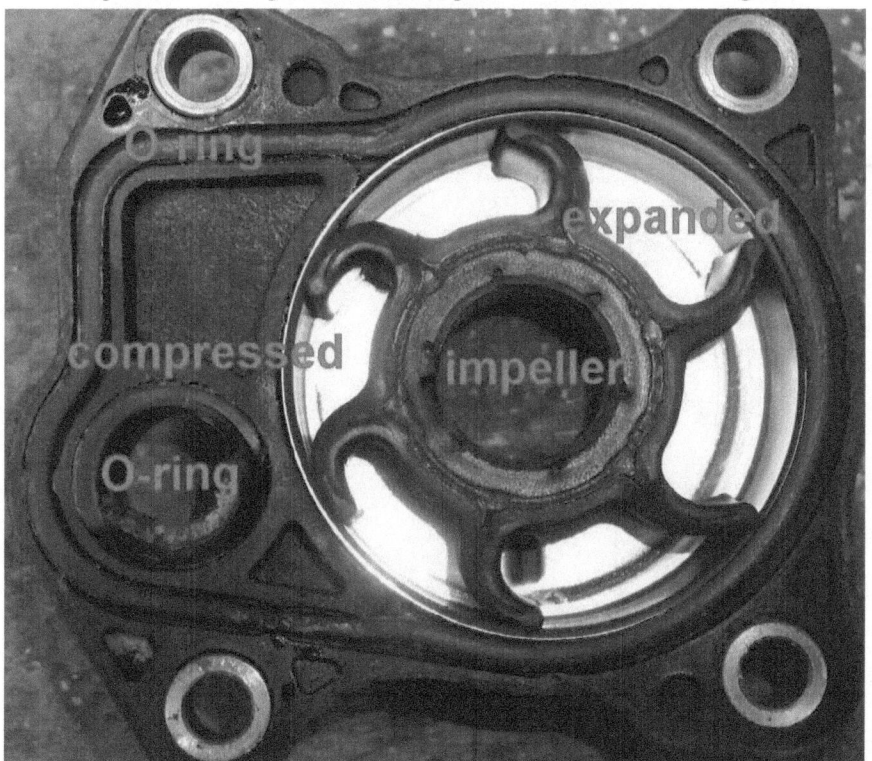

Figure 95. Impeller Distortion Produces Pumping Effect

While the diameter of these pumps does vary, large diameters are avoided. Rather, a thicker (along the drive shaft) impeller is used to increase flow. The surface speed increases with diameter (at the same rpm) so that rate of wear increases with diameter and lifetime decreases with diameter.

A typical outboard lower unit in shown in this next figure. The stainless steel plate forms the bottom surface of the water pump.

Figure 96. Typical Outboard Lower Unit (Water Pump Removed)

Water enters through the curved slot depicted just above the drive shaft. Of course, the water inlet must be kept submerged for such a pump to be effective. This is not always possible for some designs.

A side view of a typical outboard lower unit (gear case) is shown in this next figure:

Figure 97. Racing Outboard Lower Unit (Gear Case)

Depending on the position of the motor, the water intake just above the gear case may not be under water. The positioning shown below is such that half of the propeller is above the waterline. Instead of being on the motor, the water intake is on the hull.

Figure 98. Rear View of Engine Showing Vertical Positioning

While half the propeller may be out of the water at any part of the rotation, the hull will always be.

A side view shows the waterline, angle of the engine and prop shaft, as well as the hose that delivers water to the engine from the hull scoop.

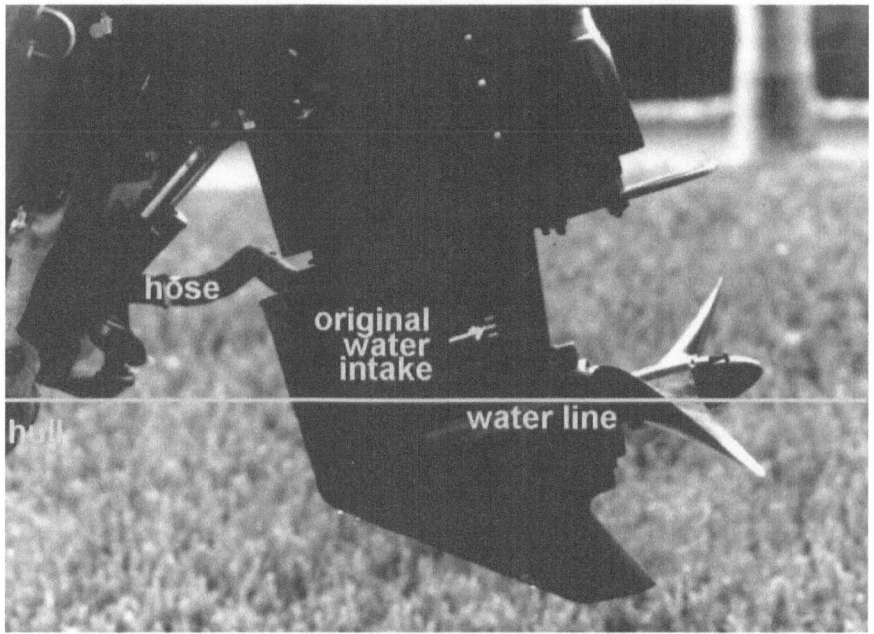

Figure 99. Side View of Engine Showing Vertical Positioning

Diagonal hydraulic ram tubes control engine angle. Some engines also have a vertical ram tube to directly control variable vertical positioning.

Hull scoop and connecting hose can be seen in the picture below:

Figure 100. Hull Scoop and Water Hose

The hydraulic pump can also be seen in the figure above.

The scoop is aligned with the hull and positioned to capture water.

Figure 101. Hull Water Scoop

This same sort of scoop is used on many inboard systems.

The final outcome of this elaborate modification was quite satisfactory:

Figure 102. Rooster Tail

Appendix A. Using the Excel Solver

The Solver function that comes with Microsoft® Excel® is a very useful tool, which can find solutions to various nonlinear equations. It is used in several example in this text. In order to use this tool, it must first be installed. This is done with from the Tools menu, as illustrated below:

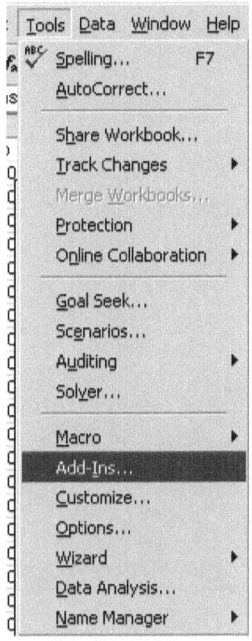

If you are using a newer version of Excel, go to the main menu, then Options, then Add-Ins, then press the button near the bottom that says G̲o... You must then check the box next to Solver.

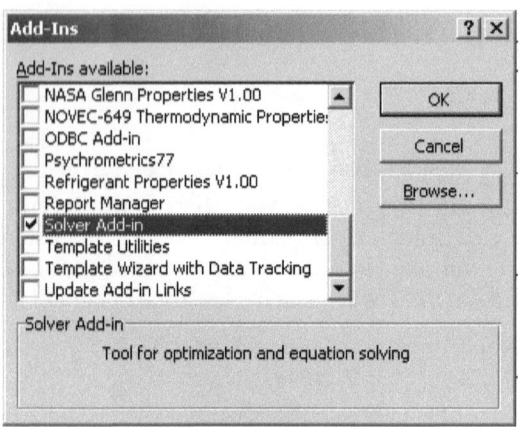

To bring up the Solver, set up the problem, and then solve it, this is also done from the Tools menu, as illustrated below:

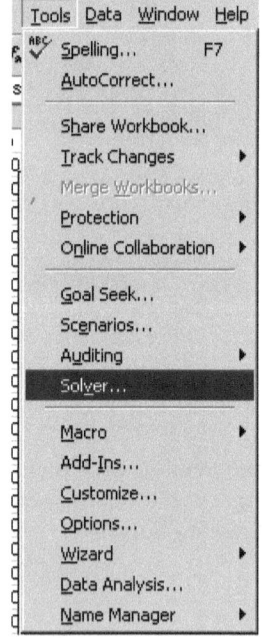

The setup dialog is shown below:

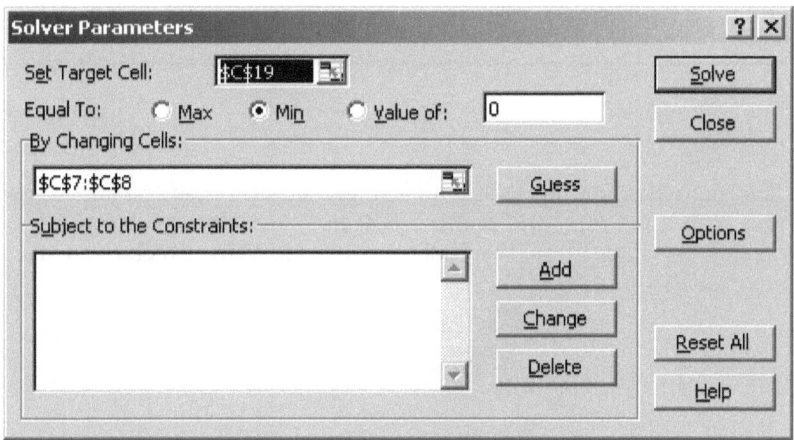

You can also use the Solver inside a macro to solve a repetitive problem, as in the prop selection spreadsheet (boat.xls) and also the fan selection spreadsheet (fan.xls). This is accomplished by inserting a call into the VBA macro, as highlighted below:

```
Option Explicit
Private Sub CommandButton1_Click()
   Dim p As Integer, i As Integer, j As Integer, pitch As
      Double, diameter As Double, d As Integer
   Range("E2:Q364").Select
   Selection.ClearContents
   Range("S2:AC34").Select
   Selection.ClearContents
   Range("S37:AC69").Select
   Selection.ClearContents
   i = 1
   For d = 20 To 30
      diameter = d / 2
      Range("C4").Value = diameter
      For p = 32 To 64
         pitch = p / 2
         Range("C5").Value = pitch
         Range("C6").Value = 1 / 9
         Range("C7").Value = 5000
         Range("C8").Value = 40
         Calculate
         SolverSolve UserFinish:=True
         i = i + 1
         For j = 1 To 13
            Cells(i, j + 4).Value = Cells(3 + j, 3).Value
         Next j
         rpm = Range("C7").Value
         If (rpm >= 5200 And rpm <= 5800) Then
```

```
          Cells(p - 30, d - 1).Value = Range("C7").Value
          Cells(p + 5, d - 1).Value = Range("C8").Value
       End If
     Next p
   Next d
End Sub
```

The first time you do this, you will get an error stating that this is not defined. You must then also add the Solver to the VBA part of Excel as a reference. This is done from the Tools menu inside VBA, which is activated by pressing Alt-F11.

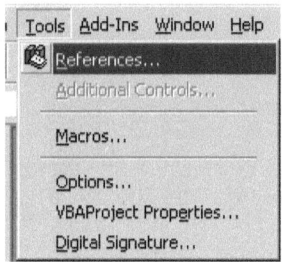

Once in the list of available references, check the box next to Solver:

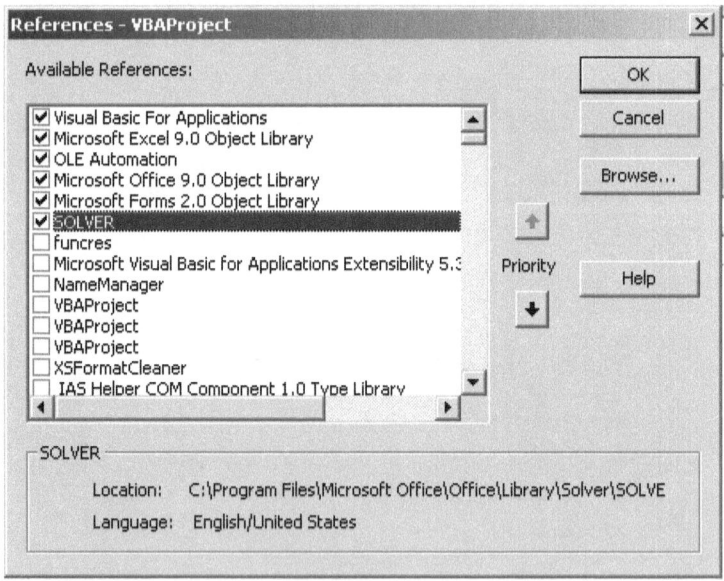

Appendix B. Reading Tecplot™ Data Files

The file extension doesn't mean anything to Tecplot™, only the headers within the file. For our purposes here, we will consider only two types: a simple list and a 2D table, both stored in ASCII point form (for example, see turbo1.dat). The first header looks like:

```
TITLE = "Garrett GTX4594R Turbocharger"
VARIABLES = "Air Flow [lb/min]"
"Pressure Ratio"
"Efficiency"
ZONE T="curves"
 I=702, J=1, K=1,F=POINT
DT=(SINGLE SINGLE SINGLE )
```

The first zone is a simple list of 702 scattered points (x,y,z). The data follow in no particular order. The second header looks like:

```
ZONE T="map"
 I=91, J=81, K=1,F=POINT
DT=(SINGLE SINGLE SINGLE )
```

The second zone is a 2D map having dimension 91x81. The data follow in row (that is, I-J). The map data can be read in with the following loop (for example, see turbo.c):

```
for(z=y=0;y<t.ny;y++)
  for(x=0;x<t.nx;x++,z++)
    if(fgets(bufr,sizeof(bufr),fp))
    sscanf(bufr,"%lf%*[ ,\t]%lf%*[
,\t]%lf",t.x+x,t.y+y,t.z+z);
```

One of the many conveniences of the C programming language is the sscanf statement above, which contains %*[,\t]. This compact notation means: read space, comma, or tab and discard the result. The space, comma, tab is inside square brackets [] and the asterisk following the percent sign means discard.

Appendix C. 2D Table Interpolation

Bivariate (i.e., 2D) linear interpolation can be quite useful and is simple to implement, especially when the data is stored in compact form. After locating the correct indices, you interpolate first over x and then over y, as shown below (for example, see turbo.c).

```
typedef struct{int nx,ny;double*x,*y,*z;}TABLE2D;
double Interpolate2D(TABLE2D t,double x,double y)
   {
   int i,j;
   double x1,x2,y1,y2,z11,z12,z21,z22,za,zb;
   for(j=0;j<t.nx-1;j++)
     if(x<t.x[j+1])
       break;
   x1=t.x[j];
   x2=t.x[j+1];
   for(i=0;i<t.ny-1;i++)
     if(y<t.y[i+1])
       break;
   y1=t.y[i];
   y2=t.y[i+1];
   z11=t.z[t.nx*i+j];
   z12=t.z[t.nx*i+j+1];
   z21=t.z[t.nx*(i+1)+j];
   z22=t.z[t.nx*(i+1)+j+1];
   za=z11+(z12-z11)*(x-x1)/(x2-x1);
   zb=z21+(z22-z21)*(x-x1)/(x2-x1);
   return(za+(zb-za)*(y-y1)/(y2-y1));
   }
```

Appendix D. Solving Pump Curves

Pump curves are most often supplied by the manufacturer. These can be readily digitized using the free tool at the web site listed in the Forward. Once digitized, curve fits can be obtained using another free tool at the same location. There is always some pressure drop or elevation change associated with any flow and so this necessitates having a pump. The head loss for a system is usually calculated and may also be available as a curve. It can be measured for an existing system if no curve is available.

Once you have the pump curve and the system loss curve, the finding operating point become a problem of locating the intersection of the two. The pump head will be falling with increasing flow and the loss curve will be rising, so that an intersection is inevitable. The simplest and most reliable way of finding the intersection is a bisection search because the lowest and highest possible values are always known and also that the solution lies somewhere between the two. The following VBA macro (see pump1.xls, pump2.xls, and pump3.xls) accomplishes this:

```
Function Flow(pumps As Integer) As Double
  Dim iter As Integer, F1 As Double, F2 As Double
  F1 = 0
  F2 = 50000
  For iter = 1 To 32
    Flow = (F1 + F2) / 2
    If (PumpHead(Flow / pumps) > SystemHead(Flow)) Then
      F1 = Flow
    Else
      F2 = Flow
    End If
  Next iter
End Function
```

also by D. James Benton

3D Articulation: Using OpenGL, ISBN-9798596362480, Amazon, 2021 (book 3 in the 3D series).

3D Models in Motion Using OpenGL, ISBN-9798652987701, Amazon, 2020 (book 2 in the 3D series.

3D Rendering in Windows: How to display three-dimensional objects in Windows with and without OpenGL, ISBN-9781520339610, Amazon, 2016 (book 1 in the 3D series).

A Synergy of Short Stories: The whole may be greater than the sum of the parts, ISBN-9781520340319, Amazon, 2016.

Azeotropes: Behavior and Application, ISBN-9798609748997, Amazon, 2020.

bat-Elohim: Book 3 in the Little Star Trilogy, ISBN-9781686148682, Amazon, 2019.

Boilers: Performance and Testing, ISBN: 9798789062517, Amazon 2021.

Combined 3D Rendering Series: 3D Rendering in Windows®, 3D Models in Motion, and 3D Articulation, ISBN-9798484417032, Amazon, 2021.

Complex Variables: Practical Applications, ISBN-9781794250437, Amazon, 2019.

Compression & Encryption: Algorithms & Software, ISBN-9781081008826, Amazon, 2019.

Computational Fluid Dynamics: an Overview of Methods, ISBN-9781672393775, Amazon, 2019.

Computer Simulation of Power Systems: Programming Strategies and Practical Examples, ISBN-9781696218184, Amazon, 2019.

Contaminant Transport: A Numerical Approach, ISBN-9798461733216, Amazon, 2021.

CPUnleashed! Tapping Processor Speed, ISBN-9798421420361, Amazon, 2022.

Curve-Fitting: The Science and Art of Approximation, ISBN-9781520339542, Amazon, 2016.

Death by Tie: It was the best of ties. It was the worst of ties. It's what got him killed., ISBN-9798398745931, Amazon, 2023.

Differential Equations: Numerical Methods for Solving, ISBN-9781983004162, Amazon, 2018.

Equations of State: A Graphical Comparison, ISBN-9798843139520, Amazon, 2022.

Evaporative Cooling: The Science of Beating the Heat, ISBN-9781520913346, Amazon, 2017.

Forecasting: Extrapolation and Projection, ISBN-9798394019494, Amazon 2023.

Heat Engines: Thermodynamics, Cycles, & Performance Curves, ISBN-9798486886836, Amazon, 2021.

Heat Exchangers: Performance Prediction & Evaluation, ISBN-9781973589327, Amazon, 2017.

Heat Recovery Steam Generators: Thermal Design and Testing, ISBN-9781691029365, Amazon, 2019.

Heat Transfer: Heat Exchangers, Heat Recovery Steam Generators, & Cooling Towers, ISBN-9798487417831, Amazon, 2021.

Heat Transfer Examples: Practical Problems Solved, ISBN-9798390610763, Amazon, 2023.

The Kick-Start Murders: Visualize revenge, ISBN-9798759083375, Amazon, 2021.

Jamie2: Innocence is easily lost and cannot be restored, ISBN-9781520339375, Amazon, 2016-18.

Kyle Cooper Mysteries: Kick Start, Monte Carlo, and Waterfront Murders, ISBN-9798829365943, Amazon, 2022.

The Last Seraph: Sequel to Little Star, ISBN-9781726802253, Amazon, 2018.

Little Star: God doesn't do things the way we expect Him to. He's better than that! ISBN-9781520338903, Amazon, 2015-17.

Living Math: Seeing mathematics in every day life (and appreciating it more too), ISBN-9781520336992, Amazon, 2016.

Lost Cause: If only history could be changed..., ISBN-9781521173770, Amazon, 2017.

Mass Transfer: Diffusion & Convection, ISBN-9798702403106, Amazon, 2021.

Mill Town Destiny: The Hand of Providence brought them together to rescue the mill, the town, and each other, ISBN-9781520864679, Amazon, 2017.

Monte Carlo Murders: Who Killed Who and Why, ISBN-9798829341848, Amazon, 2022.

Monte Carlo Simulation: The Art of Random Process Characterization, ISBN-9781980577874, Amazon, 2018.

Nonlinear Equations: Numerical Methods for Solving, ISBN-9781717767318, Amazon, 2018.

Numerical Calculus: Differentiation and Integration, ISBN-9781980680901, Amazon, 2018.

Numerical Methods: Nonlinear Equations, Numerical Calculus, & Differential Equations, ISBN-9798486246845, Amazon, 2021.

Orthogonal Functions: The Many Uses of, ISBN-9781719876162, Amazon, 2018.

Overwhelming Evidence: A Pilgrimage, ISBN-9798515642211, Amazon, 2021.

Particle Tracking: Computational Strategies and Diverse Examples, ISBN-9781692512651, Amazon, 2019.

Plumes: Delineation & Transport, ISBN-9781702292771, Amazon, 2019.

Power Plant Performance Curves: for Testing and Dispatch, ISBN-9798640192698, Amazon, 2020.

Practical Linear Algebra: Principles & Software, ISBN-9798860910584, Amazon, 2023.

Props, Fans, & Pumps: Design & Performance, ISBN-9798645391195, Amazon, 2020.

Remediation: Contaminant Transport, Particle Tracking, & Plumes, ISBN-9798485651190, Amazon, 2021.

ROFL: Rolling on the Floor Laughing, ISBN-9781973300007, Amazon, 2017.

Seminole Rain: You don't choose destiny. It chooses you, ISBN-9798668502196, Amazon, 2020.

Septillionth: 1 in 10^{24}, ISBN-9798410762472, Amazon, 2022.

Software Development: Targeted Applications, ISBN-9798850653989, Amazon, 2023.

Software Recipes: Proven Tools, ISBN-9798815229556, Amazon, 2022.

Steam 2020: to 150 GPa and 6000 K, ISBN-9798634643830, Amazon, 2020.

Thermochemical Reactions: Numerical Solutions, ISBN-9781073417872, Amazon, 2019.

Thermodynamic and Transport Properties of Fluids, ISBN-9781092120845, Amazon, 2019.

Thermodynamic Cycles: Effective Modeling Strategies for Software Development, ISBN-9781070934372, Amazon, 2019.

Thermodynamics - Theory & Practice: The science of energy and power, ISBN-9781520339795, Amazon, 2016.

Version-Independent Programming: Code Development Guidelines for the Windows® Operating System, ISBN-9781520339146, Amazon, 2016.

The Waterfront Murders: As you sow, so shall you reap, ISBN-9798611314500, Amazon, 2020.

Weather Data: Where To Get It and How To Process It, ISBN-9798868037894, Amazon, 2023.